A DIGEST OF DIGESTION

A Digest of
DIGESTION
SECOND EDITION

Horace W. Davenport, D.Sc.
William Beaumont Professor of Physiology
The University of Michigan

YEAR BOOK MEDICAL PUBLISHERS, INC.
Chicago • London

Reprinted, February 1980

Library of Congress Catalog Card Number: 78-53202
International Standard Book Number: 0-8151-2326-4

PREFACE TO
THE SECOND EDITION

Students are not learning gastrointestinal physiology to the same extent that they are learning cardiovascular, renal and respiratory physiology. One reason is the absence of a professional gastrointestinal physiologist in most physiology departments, and another is that students do not have an appropriate text. I intended my *Physiology of the Digestive Tract,* also published by Year Book Medical Publishers, to be an introductory text, but it turned out to contain more detail than a student first approaching the subject could handle in the few days allotted to gastroenterology in most current curricula. I have tried again to write a book for the beginning student. Once he has been guided in a rapid excursion through the alimentary canal, he can return for a more leisurely view of its sights with my larger book in hand.

In the preface to the first edition of this book I said, rather pompously, that references are for professors and advanced students, and I omitted them. Sure enough, professors protested their absence, and in this edition I supply some. Although the chapters in the *Handbook of Physiology: The Alimentary Canal,* edited by Charles F. Code and published by the American Physiological Society, Washington, D.C., 1968, are 15 years old, they will always be the starting place for a search of the literature. At the end of each chapter I simply list the corresponding chapters in the *Handbook.* Other references are to authoritative reviews which appear nearly monthly in *Gastroenterology* and *Gut,* to the latest paper in an important series or to symposium volumes. The only reference students ever ask me for is one on foreign objects in the rectum. I have supplied that in the appropriate place.

Successive editions of textbooks invariably develop congestive failure. The increased bulk of this book as compared with the first edition results chiefly from my response to criticisms from professors and students alike. I hope that the increase is a sign of growth from childhood to vigorous maturity and that pathological retention of words will not occur until a later edition. In the meantime, I am grateful to all who have given me help, especially to three preeminent gastroenterologists: Charles F. Code, Morton I. Grossman and Alan F. Hofmann.

H.W.D.

PREFACE TO
THE FIRST EDITION

Students are not learning gastrointestinal physiology to the same extent that they are learning cardiovascular, renal and respiratory physiology, and one reason is that they do not have an adequate text.

I intended my *Physiology of the Digestive Tract*, also published by Year Book Medical Publishers, to be an introductory text, but it turned out to contain more detail than students need or can handle during the few days allotted to gastroenterology in most current curricula. In writing this book, I have tried again to meet the needs of the beginning student.

References are for advanced students and professors, and accordingly there are no references in this book. Students who wish to know more than is given here can begin by reading my other book and by going to the library.

H. W. D.

TABLE OF CONTENTS

1. CHEWING AND THE SECRETION OF SALIVA

The amount that food is chewed depends upon the nature of the food and upon habit, and it has little effect upon the subsequent processes of digestion.

Although chewing movements can be voluntary, most chewing during a meal is a rhythmic reflex stimulated by pressure of food against the gums, teeth, hard palate and tongue. This pressure causes relaxation of the muscles holding the jaw closed against gravity, and the subsequent partial opening of the mouth decreases the strength of stimuli exerted by the food. Rebound contraction of the jaw muscles follows, and the cycle is repeated about once a second.

Most persons chew on one side of the mouth at a time.

When the contents of the mouth have been sufficiently divided, movement of the tip of the tongue separates a small bolus from the rest and brings it to the midline. Then, with the forepart of the tongue pressed firmly against the teeth and with the mouth closed, the tongue propels the bolus backward to the oral pharynx, where pressure of the bolus against receptor endings stimulates the act of swallowing.

Presence of food in the mouth and the act of chewing both stimulate secretion of saliva, and secretion is faster the larger the bite and the stronger the effort of chewing. Saliva dissolves sapid compounds and makes them available for taste; in turn, substances arousing the sensations of taste and smell stimulate secretion of saliva. Lemon drops are powerful stimuli, and when they are being sucked, flow of saliva from three pairs of glands and from numerous small buccal glands may be as high as 4 ml a minute. In man, conditioned reflexes resulting in salivary secretion are weak, and the sudden awareness of saliva already present in the mouth is probably responsible for the impression that the mouth waters at the sight or thought of food.

1

When food is not being eaten, saliva keeps the mouth wet and facilitates speech. Decreased secretion during dehydration makes the mouth dry and contributes to the sensation of thirst.

Secretion of saliva is entirely under control of autonomic nerves which innervate the glands. There is no hormonal control of salivary secretion. Although adrenergic sympathetic stimulation evokes some secretion from the submaxillary glands, the major stimuli for secretion by all glands come through parasympathetic fibers. These are cholinergic, and administration of atropine diminishes salivary secretion and makes the mouth dry.

Human saliva is a hypotonic solution whose chief cations are Na^+ and K^+ (Fig 1−1). In man, the potassium concentration at maximal rate of secretion is almost equal to that of plasma, and the sodium concentration is about three-fourths that of plasma. The anions are Cl^- and HCO_3^-, and the bicarbonate concentration is greater than that of plasma. Bicarbonate of saliva neutralizes acids of food, and because it neutralizes acids produced by bacteria in the mouth it helps to prevent caries.

In the rabbit parotid gland blood flows through two capillary beds in parallel, one supplying the intralobular ducts and one supplying the acini (Fig 1−2). In some other species, some blood may flow in a countercurrent direction, first through capillaries supplying the ducts and then to capillaries serving the acini. The arrangement in human salivary glands is not known for certain. When salivary glands are stimulated to secrete, their blood flow increases greatly. Vasodilatation is caused rapidly by acetylcholine released from parasympathetic nerve endings and slowly by vasodilator kinins formed by action of the enzyme kallikrein which is present in the apical border of cells lining the ducts.

Fluid secreted by acinar cells is isotonic with plasma, and its electrolyte composition is similar to that of an ultrafiltrate of plasma. As fluid passes down the ducts, it is modified by cells lining them. Sodium is rapidly reabsorbed, and chloride and bicarbonate passively accompany sodium. Therefore, the

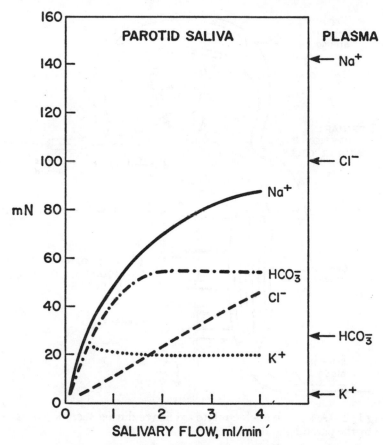

Fig 1-1.—The composition of human parotid saliva as a function of its rate of secretion. The concentrations of the ions in plasma are shown at the right-hand margin. (Adapted from Thaysen, J. H., Thorn, N. A., and Schwartz, I. L.: Am. J. Physiol. 178:155, 1954.)

number of osmotically active ions in the fluid in the ducts falls. The cells lining the ducts are almost completely impermeable to water, and water cannot diffuse from ducts to blood along an osmotic gradient. As a result, saliva is hypotonic. At low rates of salivary flow, there is more time for sodium

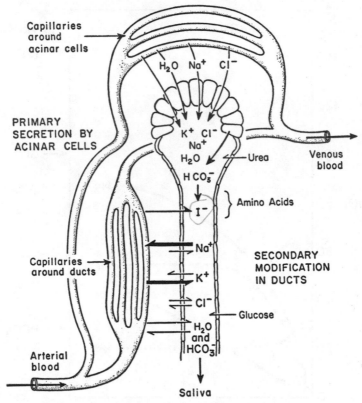

Fig 1–2.—A scheme of electrolyte exchange during secretion of saliva by the parotid gland.

reabsorption, and the osmotic pressure of saliva is lower at low rates of flow than at higher rates.

Juice secreted by the submaxillary and sublingual glands contains mucin, which is responsible for much of the lubricating action of saliva. The juice of all three pairs of glands contains ptyalin, an amylase which catalyzes the hydrolysis of the α-1,4 glucosidic linkages of starch. The enzyme is stable between pH 4 and 11, and consequently it continues to act upon starch in the part of the gastric contents that has not become acid.

SUGGESTED READINGS

Code, C. F. (ed.): *Handbook of Physiology: The Alimentary Canal* (Washington, D.C.: American Physiological Society, 1968), Chaps. 33, 34, 35, 37.

Emmelin, N., and Zotterman, Y. (eds.): *Oral Physiology* (Elmsford, N. Y.: Pergamon Press, 1972).

Fraser, P. A., and Smaje, L. H.: The organization of the salivary gland microcirculation, J. Physiol. 272:121, 1977.

Schneyer, L. H., and Schneyer, C. A. (eds.): *Secretory Mechanisms of Salivary Glands* (New York: Academic Press, 1967).

Schneyer, L. H., Young, J. A., and Schneyer, C. A.: Salivary secretion of electrolytes, Physiol. Rev. 52:720, 1972.

Shannon, I. L., Suddick, R. P., and Dowd, F. J., Jr.: *Saliva: Composition and Secretion* (Basel: Karger, 1974).

Thorn, N. A., and Petersen, O. H. (eds.): *Secretory Mechanisms of Exocrine Glands* (New York: Academic Press, 1974).

2. SWALLOWING

At rest, the esophagus is a flaccid, empty tube connecting the pharynx and the stomach. The pressure within it is the same as intrathoracic pressure (Fig 2-1). Because intrathoracic pressure is below ambient pressure, there is a pressure gradient between the mouth and the esophagus. Air would enter the esophagus from the mouth were not the junction between the pharynx and the esophagus closed by the hypopharyngeal sphincter.

Intra-abdominal pressure is greater than atmospheric pressure and still greater than intrathoracic pressure. Consequently, there is a pressure gradient from stomach to esophagus. Regurgitation of gastric contents into the esophagus is prevented by the lower esophageal sphincter. When the esophagus is at rest, this sphincter maintains a zone of pressure greater than the pressure in the stomach or esophagus.

A person begins to swallow by detaching part of the contents of the mouth, either food, drink or saliva, with the tongue and thrusting it back into the pharynx. When the bolus touches the surface of the pharynx, it initiates impulses in afferent nerves going to the swallowing center in the medulla oblongata. The swallowing center responds by sending an ordered sequence of impulses to the muscles of the pharynx, the esophagus and the stomach.

Contractions of the striated muscles of the pharynx first close the passage between mouth and nasopharynx to prevent regurgitation into the nose. Then contractions generate a wave of high pressure, averaging 200 mm Hg, which rapidly thrusts the bolus into the pharynx. The bolus often is accompanied by air which is swallowed with it. As the bolus approaches the upper esophageal, or hypopharyngeal, sphincter, muscles of the sphincter pull it apart to allow the bolus and air to enter the esophagus. Once the bolus has passed, the sphincter clo-

7

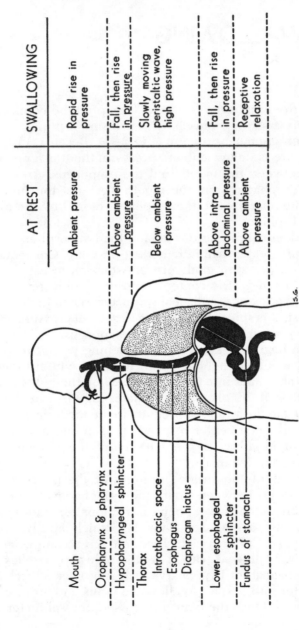

	AT REST	SWALLOWING
Mouth		
Oropharynx & pharynx	Ambient pressure	Rapid rise in pressure
Hypopharyngeal sphincter	Above ambient pressure	Fall, then rise in pressure
Thorax		
Intrathoracic space		
Esophagus	Below ambient pressure	Slowly moving peristaltic wave, high pressure
Diaphragm hiatus		
Lower esophageal sphincter	Above intra-abdominal pressure	Fall, then rise in pressure
Fundus of stomach	Above ambient pressure	Receptive relaxation

Fig 2–1.—The pharynx, esophagus and stomach. Pressures at rest and during swallowing are shown.

ses tightly at a pressure higher than its resting pressure, and it remains tightly closed for 1 to 2 seconds.

As the bolus passes the entrance to the trachea, the epiglottis folds over the glottis, the glottis closes and respiration is briefly inhibited.

Sequential, ring-like contractions of the muscle of the esophagus form a peristaltic wave which pushes the bolus toward the stomach. Pressure generated by the peristaltic wave would also push the bolus back into the pharynx, but reflux is prevented by firm closure of the hypopharyngeal sphincter during the period in which high pressure exists in the upper esophagus. The peristaltic wave moves slowly down the esophagus, taking 5 to 9 seconds for the journey (Fig 2-2).

When swallowing begins, the lower esophageal sphincter starts to relax, and it remains relaxed until the peristaltic wave at the lower end of the esophagus pushes the bolus through it into the stomach. The lower esophageal sphincter then contracts, and it maintains a pressure above the resting value for several seconds. In the meantime, the fundus of the stomach has partially relaxed to receive the bolus.

The muscles of the pharynx, hypopharyngeal sphincter and upper third of the esophagus are striated muscles innervated by motor neurons. The muscles are relaxed unless action potentials reach them through their nerves, and the vigor of their contraction is determined by the frequency of the impulses reaching them. They contract and relax in the pattern necessary to generate a moving wave of contraction, because the swallowing center sends out a stereotyped sequence of excitatory impulses to them.

The muscle of the lower two thirds of the esophagus and of the lower esophageal sphincter is smooth muscle, and its motor nerve is the vagus. Vagal cholinergic preganglionic fibers synapse with short postganglionic fibers embedded in the wall of the esophagus. Acetylcholine liberated by the postganglionic fibers causes the esophageal muscle to contract. During swallowing, the vagal fibers to the esophagus are activated in an orderly manner from above downward, so that peristalsis

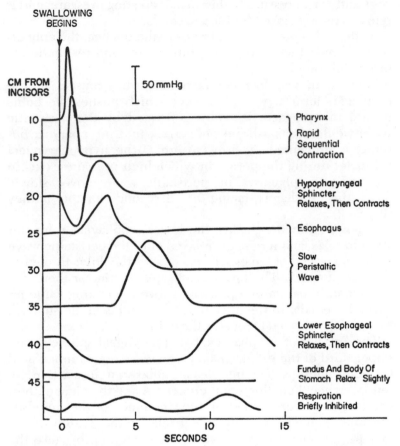

Fig 2–2. — Pressure changes occurring in the pharynx, esophagus and stomach during swallowing.

in the smooth-muscle segment of the esophagus follows smoothly the peristaltic wave in the striated muscle segment.

Although the lower esophageal sphincter is composed of smooth muscle in continuity with the rest of the esophagus, its behavior is different. Between swallows, contraction of muscle in the sphincter, independent of action of nerves, maintains a pressure in the sphincter 10–40 mm Hg higher than

that above or below the sphincter (Fig 2–3). Cholinergic preganglionic fibers in the vagus synapse with cholinergic postganglionic fibers in plexuses within the sphincter, and impulses in these nerves cause the sphincter to contract more. In
addition, excitatory fibers contained in postganglionic sympathetic nerves can cause the sphincter to contract by liberating
norepinephrine. On the other hand, impulses in noncholinergic, nonadrenergic fibers in the vagus make the sphincter

Fig 2–3.–The pressure barrier between the stomach and the esophagus is called the lower esophageal sphincter. A tube connected to a
pressure-measuring transducer is swallowed until its pressure-
sensing tip is in the stomach, and it is slowly withdrawn into the esophagus. Pressure in the stomach is above atmospheric pressure, but as
the tube is withdrawn, a segment of higher pressure approximately 2
cm long is found. After the tip of the tube has passed the diaphragm
and entered the esophagus, subatmospheric pressure is encountered.
In actual practice, variations in pressure caused by respiratory movements are superimposed on the pressure tracing.

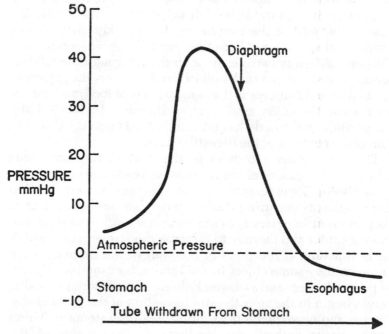

relax. The hormone gastrin (Chap. 5) increases strength of contraction of the sphincter, and the hormones secretin and progesterone decrease it. Thus, in the resting state, intrinsic contraction of the sphincter produces a barrier between esophagus and stomach, and the height of the barrier is influenced by nerve impulses and circulating hormones. When swallowing begins, the sphincter is relaxed by inhibitory impulses reaching it through vagal fibers, and at the end of swallowing, pressure in the sphincter is raised for a few seconds above resting level by excitatory impulses in vagal and sympathetic fibers.

When swallowing movements in the mouth follow one another rapidly, as in drinking, there are no peristaltic waves in the esophagus until after the last swallow; then a single normal peristaltic wave occurs.

An object falls because its specific gravity is greater than that of its surroundings. The average specific gravity of the thoracic contents is about 0.5. Consequently, in a man in the upright position, liquids, which are not held up by friction against the wall of the esophagus, fall quickly to the lower esophageal sphincter. There they wait about 5 seconds until the peristaltic wave arrives to push them through the sphincter into the stomach. On the other hand, the specific gravity of the abdominal contents is the same as that of food, and except for gas and the heavy x-ray contrast medium, barium sulfate, the position of the body has little influence upon the distribution of the contents of the digestive tract.

There are many receptors in the wall of the esophagus which send impulses along afferent nerves to the central nervous system. Some receptors respond to tension in the wall of the esophagus occurring during peristalsis, and the effect of their afferent impulses is to reinforce the outflow of the swallowing center and thereby to increase the strength of peristaltic contraction. If receptors in the wall of the esophagus are stimulated by some object in the lumen, for example, a piece of peanut butter sandwich stuck there, one or more peristaltic waves begin in the smooth-muscle section at the point of distention and move down the esophagus to the stomach. These waves occur without any preliminary movements of the mouth, and the subject is not aware that they are occurring. They are called *secondary peristaltic waves* to distinguish them from the *primary peristaltic waves* that follow swallowing movements of the mouth.

Receptors responding to tension in the esophageal wall send impulses that arouse the sensation of pain. The pain is referred to the anterior chest wall, sometimes to the posterior wall, the shoulder and the inside of the left arm. This distribution of referred pain is similar to that arising from myocardial ischemia, and it is sometimes difficult to distinguish between the esophagus and the heart as the source of the pain.

In normal adults, there is a short segment of esophagus below the diaphragm, and this segment includes part of the lower esophageal sphincter. When intra-abdominal pressure rises, so does pressure in that part of the sphincter lying below the diaphragm, simply because intra-abdominal pressure is mechanically transmitted to it as well as to the rest of the abdominal contents. This mechanism does not operate in infants who have no subdiaphragmatic segment of esophagus.

Nevertheless, reflux of gastric contents into the esophagus does occur. In normal persons, a little acid regurgitates about once an hour, but the acid is soon cleared away by eight or 10 swallows or episodes of secondary peristalsis. In some persons, resting pressure in the lower esophageal sphincter is well below normal, and in them regurgitation of acid is more frequent and prolonged. Consequently, acid bathes the mucosa of the lower esophagus for a long time. The esophagus may go into spasm, and the resulting pain, referred to the substernal area, is called heartburn.

In pregnant women, lower esophageal sphincter pressure falls to a nadir at 36 weeks, and diminished sphincter competence may allow excessive regurgitation with the result that reflux esophagitis and hearburn are often troublesome. This fall in sphincter pressure may be caused by high circulating concentration of progesterone. Oral contraceptives also reduce sphincter pressure.

In some persons, the lower esophageal sphincter fails to relax during swallowing; instead, it contracts more vigorously. Failure to relax is called *achalasia*. Passage of food from

14 A DIGEST OF DIGESTION

esophagus to stomach is greatly delayed, and the esophagus may become enormously dilated. Because a meal remains in the esophagus for a long time, food may be aspirated into the trachea, and persons with achalasia are prone to aspiration pneumonia. The underlying defect seems to be destruction of parasympathetic postganglionic fibers within the wall of the esophagus. Destruction of nerves innervating smooth-muscle cells makes the cells highly sensitive to the mediator that had been released by the nerves before they were destroyed (Cannon's law of denervation). Because the preganglionic fibers that liberate acetylcholine terminate within the wall of the esophagus, muscle cells that have become exquisitely sensitive to acetylcholine are exposed to the chemical mediator whenever the preganglionic fibers are reflexly activated. Consequently, the affected part of the esophagus goes into spasm. Sensitivity of the cells is used for differential diagnosis. A dose of a cholinergic drug that has no effect upon a normal person causes the esophagus of a patient with achalasia to contract strongly and painfully.

SUGGESTED READINGS

Code, C. F. (ed.): *Handbook of Physiology: The Alimentary Canal* (Washington, D.C.: American Physiological Society, 1968), Chaps. 90, 91, 92.

Diament, N. E., and El-Sharkawy, T. Y.: Neural control of esophageal peristalsis: A conceptual analysis, Gastroenterology 72:546, 1977.

Dodds, W. J., Hogan, W. J., Lydon, S. B., Stewart, E. T., Stef, J. J., and Arndorfer, R. C.: Quantitation of pharyngeal motor function in normal human subjects., J. Appl. Physiol. 39:692, 1975.

Dodds, W. J., Hogan, W. J., Miller, W. N., Stef, J. J., Arndorfer, R. C., and Lydon, S. B.: Effect of increased intra-abdominal pressure on lower esophageal sphincter pressure, Am. J. Dig. Dis. 20:298, 1975.

3. VOMITING

Retching and vomiting are governed by a center in the medulla oblongata which receives afferent information from two classes of receptors: (1) those in the viscera, particularly in the duodenum, which respond to constituents of the chyme such as ingested ipecac or copper salts, and (2) those in the chemoreceptor trigger zone in the area postrema of the brainstem which respond to blood-borne compounds produced in radiation sickness, in many diseases and to injected emetics such as apomorphine. In addition, vomiting is aroused by many events affecting the central nervous system. Some persons can vomit at will, and others vomit at the thought of nauseating objects. The sensation of nausea and the act of vomiting also occur in motion sickness and follow injuries producing crushing pain.

The complete act of vomiting begins with copious salivation, tachypnea, dilatation of the pupils, sweating and pallor and rapid or irregular heartbeat. These are signs of widespread autonomic discharge. The antrum of the stomach contracts with great vigor, and in it peristalsis is reversed. The duodenum goes into spasm which forces its bile-stained contents into the stomach. There may also be reversed peristalsis in the small intestine. On the other hand, the body of the stomach and the lower esophageal sphincter relax completely (Fig 3–1).

Slow, deep inspirations against a partially closed glottis produce moaning sounds and reduce intrathoracic pressure far below atmospheric pressure. At the same time, strong contractions of the abdominal muscles increase intra-abdominal pressure. The large pressure gradient from abdomen to thorax forces contents of the flaccid body of the stomach through the relaxed lower esophageal sphincter into the esophagus. Reverse peristalsis never occurs in the human esophagus. The

15

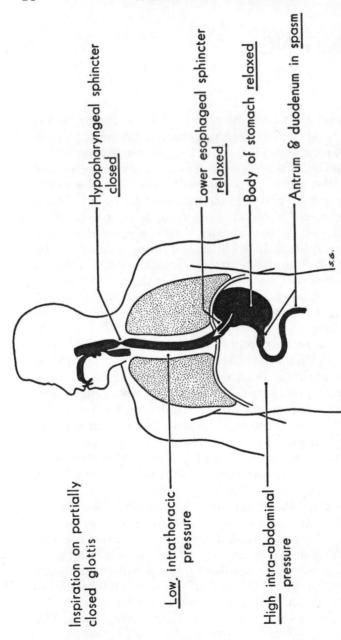

Inspiration on partially closed glottis

Hypopharyngeal sphincter closed

Lower esophageal sphincter relaxed

Body of stomach relaxed

Antrum & duodenum in spasm

Low intrathoracic pressure

High intra-abdominal pressure

Fig 3–1.— The mechanics of retching. Duodenal contents are forced into the stomach by spasm of the duodenum. The antrum is also in spasm, but the body of the stomach and the lower esophageal sphincter are relaxed. Inspiration against a partially closed glottis lowers intrathoracic pressure, and contraction of abdominal muscles raises intra-abdominal pressure. The pressure gradient from abdomen to thorax forces contents of the body of the stomach into the esophagus. The hypopharyngeal sphincter remains closed, and secondary peristaltic waves in the esophagus force its contents back into the stomach. The cycle may be repeated many times.

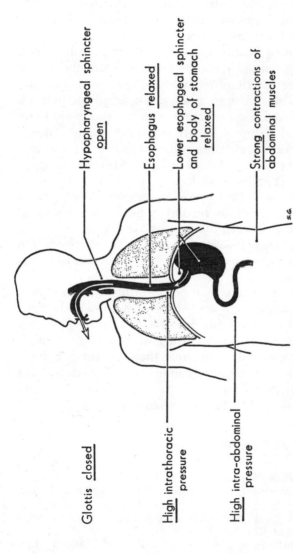

Fig 3–2.– The mechanics of vomiting. Antrum and duodenum are in spasm, but the body of the stomach and the lower esophageal sphincter are relaxed. While the esophagus is full as the result of a retch, strong contractions of the abdominal muscles force the diaphragm high into the thorax. Because the glottis is closed, intrathoracic pressure rises and forces the contents of the esophagus through the open hypopharyngeal sphincter and out the mouth. There may be reversed peristalsis in the duodenum and antrum, but there is no reversed peristalsis in the body of the stomach or in the esophagus.

Labels in figure:
- Glottis closed
- Hypopharyngeal sphincter open
- Esophagus relaxed
- Lower esophageal sphincter and body of stomach relaxed
- Strong contractions of abdominal muscles
- High intrathoracic pressure
- High intra-abdominal pressure

hypopharyngeal sphincter remains closed, and no gastric contents enter the mouth. Distention of the esophagus induces secondary peristalsis, and gastric contents are swept back into the stomach. This is the act of retching, and it may be repeated many times before vomiting occurs.

The act of vomiting builds upon a developed retch. Strong contractions of abdominal muscles increase intra-abdominal pressure and force the diaphragm high into the thorax. Because the glottis is closed, there is a sudden, large rise in intrathoracic pressure. At the same time, the larynx and hyoid bone are drawn forward to open the hypopharyngeal sphincter. Then the high intrathoracic pressure, which may be 100 mm Hg above ambient pressure, forces the contents of the esophagus out the mouth (Fig 3–2). Abdominal muscles relax, and if the esophagus has not been completely emptied, its contents are returned to the stomach by secondary peristalsis. A second cycle of filling and emptying the esophagus may begin again, culminating in further expulsion of vomitus. Throughout, the body of the stomach remains flaccid and does not contribute to expulsion of its contents.

In the young infant, there may be no intra-abdominal segment of the esophagus, and an increase in intra-abdominal pressure does not reinforce the lower esophageal sphincter. Consequently, regurgitation occurring after a large meal is more like bubbling over of a fumarole than vomiting. Babies, however, can vomit in the adult manner, and they often vomit when they are enraged. That they eventually lose the ability to vomit voluntarily is a great relief to parents.

SUGGESTED READINGS

Code, C. F. (ed.): *Handbook of Physiology: The Alimentary Canal* (Washington, D.C.: American Physiological Society, 1968), Chaps. 90, 92, 108.

Borison, H. L., and Wang, S. C.: Physiology and pharmacology of vomiting, Pharmacol. Rev. 5:193, 1953.

Fordtran, J. S.: Vomiting, in Sleisinger, M. H., and Fordtran, J. S. (eds.): *Gastrointestinal Disease* (Philadelphia: W. B. Saunders Company, 1973).

4. MUSCLE AND NERVES OF THE GUT

Smooth muscle of the gut is different in structure and function from striated muscle. The two kinds of muscle are contrasted in Table 4-1.

Individual smooth-muscle fibers are very small, and they are organized in bundles of about 200 cells in cross-section. The cells are connected one with the other by gap junctions through which current can flow from one cell to the other. The result is that each bundle is in effect a physiological syncytium. The bundles are arranged in layers. A thin longitudinal layer runs lengthwise along the second half of the esophagus, the whole of the stomach and the small intestine. Strips of the longitudinal layer form the taeniae coli of the colon. Beneath the longitudinal layer is a thicker and stronger circular layer of bundles, each bundle roughly forming a ring around its hollow organ. Another thin layer of muscle, the muscularis mucosae, lies between the circular muscle and the mucosa.

The muscles of the pharynx, the hypopharynx and the first third of the esophagus are striated muscles, and they are totally dependent upon their efferent nerves for any function. At the other end of the gut, the external anal sphincter is also composed of striated muscle, and if its efferent nerves are destroyed, the sphincter is paralyzed. Between these two extremes lie the smooth-muscle structures of the gut, and the most important point to understand about the function of these structures is that they are provided with a nervous system of their own which is capable of executing the functions of the gut without any extrinsic innervation whatever. The extrinsic innervation this nervous system receives from the sympathetic and parasympathetic nerves modulates but does not command its activity.

Each smooth-muscle fiber has a transmembrane potential,

19

TABLE 4-1.—COMPARISON OF SMOOTH MUSCLE OF THE GUT WITH STRIATED MUSCLE

SMOOTH MUSCLE OF THE GUT	STRIATED MUSCLE
Very small fibers organized in bundles	Long fibers organized in muscle units
Muscle fibers within bundles connected by gap junctions	Fibers independent of each other except by common innervation by one motor neuron
Origin and insertion in connective tissue	Origin and insertion on bones
Innervated by fibers of nervous plexuses; a potentially independent nervous system that can act without the CNS	Innervated by motor neurons; entirely dependent on the CNS
Muscle fibers have no motor end plates; efferent fibers from plexuses liberate mediators near the cell surface	Each fiber has a motor end plate
Ganglion cells of plexuses innervated by preganglionic parasympathetic fibers and equivalent to parasympathetic postganglionic cell bodies	No plexuses
Ganglion cells of plexuses innervated by postganglionic sympathetic fibers; sympathetic activity modulates activity of ganglion cells by inhibiting on-going activity and reflexes	
Sympathetic postganglionic innervation of muscle cells exists but is of questionable importance	Circulating sympathetic mediators have minor effects on muscle cells; no sympathetic innervation of muscle cells
Muscle fiber membrane potential variable at rest	Membrane potential constant at rest
Spontaneous variations in membrane potential	None
Hyperpolarization with increased threshold	
Hypopolarization with decreased threshold	
Basic electrical rhythm = rhythmic depolarization conducted along longitudinal fiber bundles	
Electrotonic current flow between longitudinal fiber bundles and circular fiber bundles, influencing the threshold of circular fiber bundles	None

TABLE 4-1.—*Continued*

SMOOTH MUSCLE OF THE GUT	STRIATED MUSCLE
Action potentials may or may not occur during rhythmic depolarization	Action potentials follow end plate depolarization
Occurrence of action potentials strongly influenced by	No plexuses; slight effect of circulating hormones
Activity in plexuses: cholinergic output decreasing muscle fiber threshold	
Circulating hormones: epinephrine and norepinephrine increasing muscle fiber threshold; gastrin and secretin with specific effects on specific muscles	
Active tension always present; modulated by mediators and hormones; varies with resting membrane potential	No active tension without muscle action potential
Variable response to stretch: elastic lengthening plus either stress relaxation or active contraction	Elastic lengthening only
Variable response to release of stretch: elastic shortening plus either active contraction or further relaxation	Elastic shortening only
Active tension increases following action potential	No active tension without muscle action potential
Very wide range of active tension; large tetanus-to-twitch ratio	Narrow range of tetanus-to-twitch ratio; large range of tension achieved by spatial and temporal summation
Slow contraction and relaxation	Rapid contraction and relaxation

positive outward and negative inward. This membrane potential, like that of striated muscle and nerve, is in part a potassium diffusion potential. The concentration of potassium is higher inside the smooth muscle cell than it is in the extracellular fluid, and the tendency of potassium ions to diffuse outward along their concentration gradient, carrying with them a positive charge, contributes to the transmembrane potential. Other ions also contribute to the transmembrane potential of smooth-muscle cells; both sodium and calcium ions tend to diffuse inward, carrying positive charges which oppose those of potassium diffusing in the opposite direction. However, the transmembrane potential of intestinal smooth-muscle cells cannot be explained entirely as the resultant of several diffusion potentials. The muscle cell membrane appears to contain an electrogenic sodium pump. This pump, using metabolic energy, pumps sodium ions out of the cell and therefore tends to make the outside of the cell positive with respect to the inside.

Variations in the activity of the electrogenic sodium pump account for much of the variability of the resting membrane potential of smooth-muscle cells. The pump is stimulated by circulating epinephrine and norepinephrine, and as it vigorously extrudes sodium ions, the transmembrane potential becomes larger. The cells are hyperpolarized and less excitable, and their tension decreases (Fig 4–1). On the other hand,

Fig 4–1.—The effect of epinephrine upon a smooth-muscle cell of the longitudinal and circular muscle bundles of the gut. Epinephrine hyperpolarizes the cell. When the transmembrane potential is greater than threshold, action potentials do not occur. A decrease in tension follows the decrease in frequency of action potentials and the increase in transmembrane potential. Norepinephrine has the same effect.

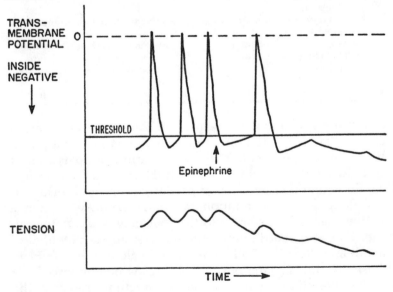

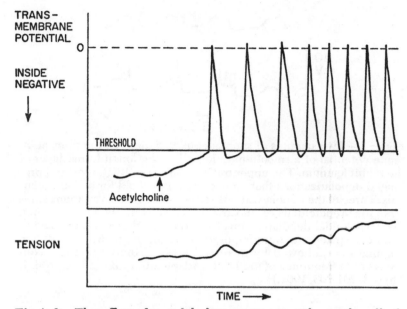

Fig 4–2.—The effect of acetylcholine upon a smooth-muscle cell of the longitudinal and circular muscle bundles of the gut. Acetylcholine causes depolarization of the cell, and when the membrane potential reaches threshold, an action potential occurs. Since acetylcholine increases the rate of depolarization, it increases the frequency of action potentials. An increase in tension follows the action potentials.

acetylcholine depolarizes the cells, decreasing their membrane potential, increasing their tension and making them more excitable (Fig 4–2).

When a smooth-muscle cell is depolarized to a certain extent, its membrane has an action potential. Like the action potentials of striated muscle cells and neurons, this is an abrupt depolarization which spreads over the whole of the cell's membrane and which is followed by repolarization.

Smooth-muscle cells have three kinds of variations in their resting membrane potential (Fig 4–3).

1. The membrane potential may slowly increase, in which case the cell is hyperpolarized and less excitable, or the

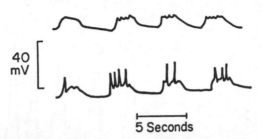

Fig 4–3.—Two samples of spontaneous variations in the transmembrane potential of a smooth-muscle cell in the longitudinal layer of the rabbit jejunum. The upper tracing shows the BER, repeated prolonged depolarizations that do not reach threshold for action potentials. Three of the depolarizations show small further variations similar to the prepotentials of sinoatrial nodal cells. In the lower tracing, the prepotential depolarizations have reached threshold, and action potential spikes are superimposed upon the BER. Because action potentials are followed by increased tension, the muscle cell contracts at the frequency of the BER. (Adapted from Bortoff, A.: Am. J. Physiol. 201:203, 1961.)

membrane potential may slowly decrease, in which case the cell is hypopolarized and more excitable. These changes may be spontaneous in the sense that their cause is not identified, or they may be the effects of changes in membrane permeability and sodium pumping resulting from changes in concentration of hormones or chemical mediators in the cell's environment.

2. Some smooth-muscle cells of the gut may have rhythmic depolarizations similar to the prepotentials of the sinoatrial node of the heart. If one of these depolarizations reaches threshold, an action potential follows. The frequency of action potentials in a particular cell, therefore, depends upon (a) the frequency with which the prepotentials occur, (b) the rate at which the prepotentials move to threshold and (c) the threshold at the moment.

3. Some cells of the longitudinal muscle bundles (but not those of the circular muscle bundles) have spontaneous rhythmic depolarizations occurring at a frequency characteristic of the organ. In the stomach, these cells are lo-

cated high on the greater curvature, and their frequency of depolarization is approximately three depolarizations per minute. This train of spontaneous depolarizations, called the Basic Electrical Rhythm (BER), Pacesetter Potential (PSP) or Electrical Control Activity (ECA), is conducted along the longitudinal muscle bundles, and in the stomach it is responsible for establishing the frequency and the rate of progress of peristaltic waves.

Smooth-muscle cells contain myofilaments of actin and myosin, but the filaments are not arranged in the elegant array characteristic of striated muscle fibers. The intimate mechanism by which chemical energy is transformed into mechanical energy appears to be the same in both kinds of muscle. The chemical machine in smooth-muscle cells, however, is continuously active; smooth-muscle cells of the gut always have some degree of tension during life. This degree of tension is related to the membrane potential: a decrease in transmembrane potential, or hypopolarization, is associated with an increase in active tension, and an increase in transmembrane potential, or hyperpolarization, is associated with a decrease in tension, or relaxation. In addition, an action potential is followed by a large increase in active tension. Changes in active tension are slow, and fusion of contractions following action potentials occurs readily. Tetanus of smooth muscle occurs at a low frequency of action potentials. The range of tension development by gut smooth muscle, from almost complete relaxation to maximal contraction, is very wide.

An action potential is conducted over the membrane of a single cell, but it cannot be conducted from a single cell to another single adjacent cell. When an action potential occurs in a cell's membrane, the transmembrane potential becomes zero, or it overshoots and reverses its sign. Consequently, there is a potential difference between the depolarized cell and its neighboring, undepolarized cell, and current flows from the positively charged surface of the undepolarized cell through the extracellular fluid to the depolarized cell. The circuit is completed by flow of current from the interior of the depolarized cell to the interior of the adjacent cell through the gap junction connecting the two cells. If the amount of current flowing is sufficient to depolarize the neighboring cell to threshold, an action potential will occur in the neighboring cell, and the action potential will, in effect, be conducted from one cell to another. The difficulty in smooth muscle of the gut is that each cell is very small, and therefore the surface-to-volume ratio is very large. A completely depolarized cell acts as a current sink with respect to its neighbors, but because the volume of the cell is small, not enough current flows to bring the membrane of the neighboring cells to threshold (Fig 4–4). To circumvent this problem, smooth-muscle cells in the gut

Fig 4–4.– Current flow between a depolarized smooth-muscle cell of the gut and an adjacent polarized cell. Current flows in the external circuit from the surface of the polarized cell to the depolarized cell, here represented as having a reversal or overshoot of its membrane potential. The circuit is completed by flow of current from the depolarized cell to its neighbor through their gap junction. The cells are so small that current flow is not enough to bring the polarized cell to threshold and to initiate an action potential. When about 1,000 smooth-muscle cells in a bundle are simultaneously depolarized, however, current flow is sufficient to bring neighboring cells to threshold, and action potentials are propagated along the bundle.

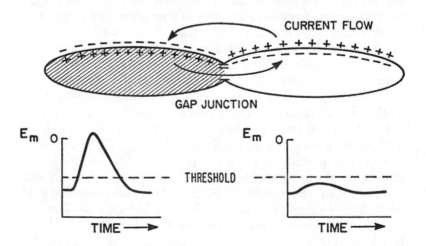

are organized in bundles. If a thousand cells, a group five cells long and two hundred cells in cross-section, are simultaneously depolarized, they form a large current sink. Enough current flows from the neighbors of these thousand cells so that the neighbors themselves reach threshold and experience action potentials. Excitation is thus conducted along bundles, and the cells of the bundles contract as a unit.

As the conducted wave of partial depolarization known as the BER passes slowly along bundles of the longitudinal muscle, it acts as a current sink for the underlying circular muscle. Therefore, there is an electrotonic flow of current from successive rings of circular muscle to the longitudinal muscle. This current flow tends to depolarize the bundles of circular muscle and bring them toward their threshold for action potentials. In some parts of the gut, small bundles of fibers connect the longitudinal with the circular layer, and these carry excitation from the one to the other. Whether or not the circular muscle bundles reach threshold, have action potentials and contract depends on their threshold, or excitability, at the moment, and their threshold at the moment depends in turn upon the amount of acetylcholine liberated near the muscle fiber bundles by nerves of the local plexuses.

There are two major networks of nerve fibers forming the plexuses of the gut: (1) the myenteric plexus between the two muscle layers and (2) the submucous plexus lying between the mucosa and the circular muscle layer. Together these form a complete and competent nervous system, and they are the evolutionary descendants of the primitive nervous system in the wall of such animals as the sea anemone.

The plexuses contain cells whose afferent fibers end in receptors in the wall of the gut or in the mucosa (Fig 4–5). The receptors may be chemoreceptors sensitive to components of the chyme such as hydrogen ions or polypeptides, or they may be mechanoreceptors sensitive to stretch or tension. Efferent fibers of the receptor cells synapse with cell processes and cell bodies within the plexuses to form an integrating nervous system. Efferent fibers leave cells of the plexuses to reach muscle cells of the longitudinal and circular layers and of the muscu-

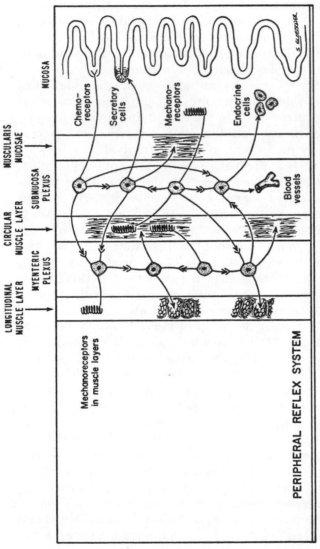

Fig 4–5. — The peripheral reflex system of the gut. The intrinsic plexuses form a nervous system capable of integrative action without the aid of the extrinsic nerves. Chemo- and mechanoreceptors send impulses to cell bodies of the plexuses. The cell bodies have efferent processes that synapse with many other cells of the plexuses, and efferent fibers from the latter cells stimulate or inhibit muscle cells in the muscle layers and in the blood vessels, stimulate secretory cells of the mucosa or regulate release of hormones from endocrine cells.

laris mucosae. When the efferent fibers liberate acetylcholine, the fibers are excitatory. By liberating other mediators, some other fibers may be inhibitory. Efferent fibers also reach secretory cells of the mucosa, and their mediators stimulate secretion of juices into the lumen or hormones into the blood. Distention of the stomach by neutral protein digestion products provides an example of the action of this intrinsic nervous system. Distention activates stretch receptors and, through a cholinergic reflex containing at least one synapse, stimulates release of gastrin by cells in the antrum and secretion of acid by oxyntic cells in the body of the stomach. The polypeptides stimulate chemoreceptors and, through a similar cholinergic reflex, they too stimulate release of gastrin and secretion of acid.

Extrinsic innervation arrives by two pathways: (1) parasympathetic preganglionic fibers running in the vagus nerve (Fig 4–6) and (2) sympathetic postganglionic fibers running along blood vessels (Fig 4–7).

Parasympathetic preganglionic fibers end on ganglion cells of the intrinsic plexuses of the gut, and therefore the fibers of the ganglion cells can be called postganglionic fibers. Such nomenclature is misleading and meaningless, however. In all parasympathetic systems outside the gut, the postganglionic fibers have no function except that dictated by their preganglionic innervation. In the gut, the ganglion cells and their processes are part of an autonomous nervous system capable of acting without parasympathetic innervation. The function of the parasympathetic innervation is to modulate, not absolutely control, the activity of the nervous system contained in the plexuses.

Most of the parasympathetic fibers to the plexuses of the gut are cholinergic and excitatory. An example of their action is seen during hunger. Then the glucose-sensitive cells of the hypothalamus excite the cells of the vagal nucleus, and action potentials run along vagal fibers to the stomach. Ganglion cells are excited, and their efferent processes liberate acetylcholine near muscle fibers of both the longitudinal and circular muscle bundles. In the longitudinal layer, the frequency

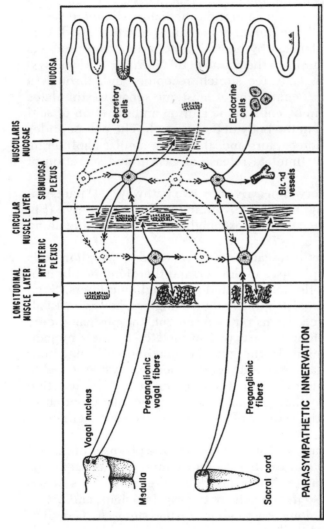

Fig 4–6.—The parasympathetic innervation of the gut. Parasympathetic preganglionic fibers synapse with cells of the intrinsic plexuses and regulate the activity of the plexuses. In many instances, the parasympathetic fibers are cholinergic and excitatory, and their action increases motility, secretion and release of hormones. In other important instances, they are noncholinergic and inhibitory, and they mediate such reflexes as relaxation of the lower esophageal sphincter and the stomach during swallowing.

PARASYMPATHETIC INNERVATION

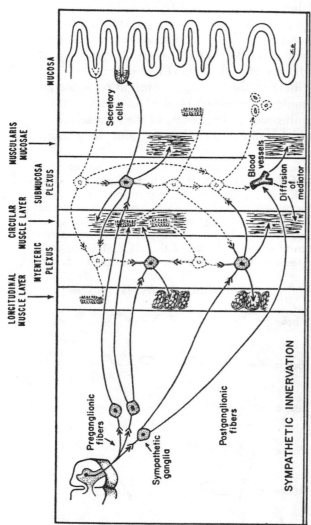

Fig 4-7.—The sympathetic innervation of the gut. Postganglionic sympathetic fibers synapse with the cells in the intrinsic plexuses and regulate the activity of the plexuses. In most instances, they are adrenergic and inhibitory, and their action is to inhibit ongoing activity in the plexuses. Many sympathetic adrenergic fibers innervate the smooth muscle of the blood vessels, causing vasoconstriction, and norepinephrine released at the blood vessels inhibits nearby cells of the muscle layers. Some adrenergic sympathetic fibers end near cells of the muscle layers. Action of the sympathetic nerves to the gut is backed up by epinephrine and norepinephrine released by the adrenal medulla.

and velocity of the BER are slightly increased. In the circular layer, the excitability of the muscle fibers is raised; electrotonic flow of current between them and the longitudinal layer brings them to threshold. They have action potentials and contract. A ring of contraction, the peristaltic wave, sweeps over the stomach, following the BER as it traverses from high on the gastric wall to the pyloric sphincter (Fig 4–8).

Not all efferent fibers in the vagus are excitatory; some are inhibitory. An example is furnished by the innervation of the lower esophageal sphincter and the body of the stomach. Both of these structures relax during swallowing, and the efferent

Fig 4–8.— The origin and control of peristaltic waves in the antrum of the stomach. A pair of electrodes, 6.8 cm apart, in the longitudinal muscle records the BER at a frequency of 3 per minute and moving from the body of the stomach toward the pyloric sphincter at a velocity of 1 cm per second. Action potentials in the circular muscle accompany the BER, and consequently a wave of peristaltic contraction sweeps over the antrum at the velocity of the BER. Two or more waves can often be seen in sequence in the antrum of the human stomach.

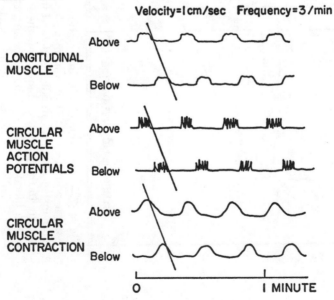

fibers mediating relaxation are in the vagus. These particular fibers are neither adrenergic nor cholinergic. The best present evidence is that the vagal preganglionic fibers liberate 5-hydroxytryptamine and that the corresponding postganglionic fibers liberate adenosine triphosphate. Other vagal fibers carry impulses that inhibit gastric secretion.

There are apparently no vasodilator fibers in the parasympathetic innervation of the gut. Vasodilation and increased blood flow do occur when excitatory impulses in the parasympathetic system enhance motility and secretion, but this is active hyperemia, the result of increased metabolism of muscle and glands.

The fact that parasympathetic impulses to the gut enhance reflex activity of the plexuses does not mean that this is the only function of parasympathetic innervation. There are reflexes whose afferent and efferent pathways are both in the vagal or sacral nerves. About 90% of the fibers in the vagus are afferent, and many of these fibers have mechanoreceptor or chemoreceptor endings in the stomach, intestine and proximal colon. Receptor endings in the distal colon send impulses to the sacral section of the spinal cord and, among other things, enhance the defecation reflex. An example of a vagally mediated reflex is the relaxation of the stomach that follows distention of the esophagus. Impulses arising in stretch receptors in the esophagus travel centrally in the vagus, and efferent impulses travel in nonadrenergic, noncholinergic vagal fibers to the stomach.

Afferent fibers in both vagal and sympathetic nerves carry impulses which arouse somatic reflexes. For example, ipecac or copper sulfate in the duodenum causes vomiting. Afferent fibers also mediate referred pain.

The sympathetic nervous system affects the gut through four pathways: (1) by adrenergic postganglionic fibers terminating on ganglion cells of the plexuses, (2) by adrenergic postganglionic fibers going to some muscle bundles, (3) by epinephrine and norepinephrine from the adrenal medulla, and (4) by adrenergic vasoconstrictor fibers to smooth muscle of the blood vessels of the gut.

Norepinephrine liberated at the surface of ganglion cells by sympathetic fibers depresses their excitability and thereby inhibits any on-going activity. Therefore, sympathetic discharge tends to inhibit reflexes mediated through the plexuses or through the vagus nerve. Thus, the peristaltic reflex stimulated by distention of the small intestine is suppressed by sympathetic impulses to ganglion cells in the wall of the intestine.

Some adrenergic postganglionic sympathetic fibers end near muscle cells of circular bundles. How important these are is unknown.

In the gut, as in the rest of the body, sympathetic influences are backed up by epinephrine and norepinephrine liberated by the adrenal medulla. These hormones stimulate vascular smooth muscle of the gut, thereby causing vasoconstriction, and they inhibit intestinal smooth muscle, thereby decreasing motility.

Adrenergic postganglionic sympathetic fibers innervate the smooth muscle of the blood vessels of the gut. When they are stimulated, there is an immediate vasoconstriction and eventually decreased intestinal motility. This decrease in motility is probably the result of diffusion of norepinephrine from the blood vessels to neighboring smooth-muscle cells of the longitudinal and circular bundles.

SUGGESTED READINGS

Code, C. F. (ed.): *Handbook of Physiology: The Alimentary Canal* (Washington, D.C.: American Physiological Society, 1968), Chaps. 80–88.

Bortoff, A.: Myogenic control of intestinal motility, Physiol. Rev. 56: 418, 1976.

Bülbring, E., Brading, A. F., Jones, A. W., and Tomita, T. (eds.): *Smooth Muscle* (Baltimore: Williams & Wilkins Co., 1970).

Furness, J. B., and Costa, M.: The adrenergic innervation of the gastrointestinal tract, Ergeb. Physiol. 69:1, 1974.

Sarna, S. K.: Gastrointestinal electrical activity: Terminology, Gastroenterology 68:1631, 1975.

Weems, W. A., and Szurszewski, J. H.: Modulation of colonic motility by peripheral neural input to neurones of the inferior mesenteric ganglion, Gastroenterology 73:273, 1977.

Wood, J. D.: Neurophysiology of Auerbach's plexus and control of intestinal motility, Physiol. Rev. 55:307, 1975.

5. HORMONES OF THE GUT

The three best understood hormones of the gut are gastrin (or the family of gastrins), cholecystokinin-pancreozymin (CCK-PZ, or simply CCK) and secretin. Their structures are given in Table 5–1, and their major actions are listed in Table 5–2. Their structures need not be memorized, but their homologies should be recognized.

The gastrins are a family of straight-chain polypeptides. *Little gastrin* is a heptadecapeptide (17 amino acids). The N-terminal acid is glutamic acid in the pyro form, and the carbonyl group at the C-terminal phenylalanine is aminated. These forms of the terminal amino acids protect the peptide from degradation by amino- and carboxypeptidases. The tyrosine in position 12 may or may not be sulfated; whether it is makes no difference to the action of the hormone. Species differences result from a single base substitution in the genetic codon. Thus, human gastrin has leucine in position 5, whereas porcine gastrin has methionine. Because these substitutions occur in the nonspecific part of the molecule, they have no physiological significance. The half-life of circulating little gastrin is about 7 minutes.

Big gastrin, which also may or may not be sulfated, consists of little gastrin to which an additional chain of 17 amino acids is added to the number 1 amino acid of little gastrin. Big gastrin has the same spectrum of actions as little gastrin, but its half-life is longer, being about 42 minutes in man. As a consequence of its longer half-life, the concentration of big gastrin in blood rises higher and remains high longer when equimolar amounts of the two gastrins are released.

Other gastrins and gastrin fragments, including a pair of *mini-gastrins* 14 amino acids long and a physiologically inactive peptide consisting of the first 13 amino acids of little gas-

TABLE 5-1.—MAJOR HORMONES OF THE GUT

LITTLE HUMAN GASTRIN	ACTIVE GROUPS
1 2 3 4 5 6 7 8 9 10 11 12 13 Pyr-Gly-Pro-Trp-Leu-Glu-Glu-Glu-Glu-Glu-Ala-Tyr-Gly- SO_3	14 15 16 17 -Trp-Met-Asp-Phe-NH$_2$

Human gastrin I is not sulfated at Tyr$_{12}$

CHOLECYSTOKININ-PANCREOZYMIN

26 27 28 29 25 more-Asp-Tyr-Met-Gly- SO_3	30 31 32 33 -Trp-Met-Asp-Phe-NH$_2$

PENTAGASTRIN (SYNTHETIC)

$C(CH_3)_3$-OCO-NH-CH$_2$-CH$_2$-CO-	-Trp-Met-Asp-Phe-NH$_2$

SECRETIN

1 2 3 4 5 6 7 8 9 10 11 12 13
His-*Ser*-Asp-*Gly*-*Thr*-*Phe*-*Thr*-*Ser*-Glu-Leu-*Ser*-Arg-Leu-

14 15 16 17 18 19 20 21 22 23 24 25
Arg-*Asp*-*Ser*-Ala-*Arg*-Leu-*Gln*-Arg-Leu-Leu-*Gln*-Gly-

26 27
Leu-Val-NH$_2$

Note: The 14 amino acids in italics occupy the same position, counting from the N-terminus, as do those in glucagon. There is no active group, the whole molecule being required.

trin, have been found in normal and pathological serum and tissues, but their significance is unknown.

In all gastrins, the last four amino acids, counting from the N-terminus, are -Trp-Met-Asp-Phe-NH$_2$. These four amino acids are the active group of the molecule. A tetrapeptide of this sequence has all the physiological properties of gastrin, although on a molar basis it is less potent than the larger molecules. The synthetic compound, pentagastrin, consists of these four amino acids linked to a substituted β-alanine, and it is

TABLE 5-2.—MAJOR PHYSIOLOGICAL ACTIONS OF GASTRIN, CHOLECYSTOKININ AND SECRETIN

GASTRIN

Increases resting pressure in lower esophageal sphincter

IMPORTANT: Stimulates acid secretion by the oxyntic cells which in turn stimulates secretion of pepsinogen by chief cells through local reflex

IMPORTANT: Increases gastric antral motility

Weakly stimulates enzyme and bicarbonate secretion by pancreas, contraction of gallbladder

IMPORTANT: Has trophic effects on gastric mucosa

CHOLECYSTOKININ

Weakly stimulates gastric secretion of acid

IMPORTANT: Competitively inhibits gastrin-stimulated secretion of acid

IMPORTANT: Strongly stimulates enzyme secretion by pancreas

Weakly stimulates secretion of bicarbonate by pancreas, BUT

IMPORTANT: Strongly potentiates effect of secretin in stimulating bicarbonate secretion by pancreas

IMPORTANT: Strongly stimulates contraction of gallbladder

Stimulates duodenal secretion and motility

Slows gastric emptying

IMPORTANT: Has trophic action on pancreas

SECRETIN

Stimulates pepsinogen secretion

IMPORTANT: Stimulates secretion of bicarbonate by pancreas and liver; synergistic with CCK

IMPORTANT: Inhibits gastrin-stimulated acid secretion

IMPORTANT: Inhibits gastric and duodenal motility

Inhibits lower esophageal sphincter

Has metabolic effects similar to those of glucagon

replacing histamine in tests of acid secretion. The longer chain of amino acids attached to the active group confers on gastrin its quantitative properties. The entire molecule is a very powerful stimulant of acid secretion, but it is only a weak stimulant of gallbladder contraction.

Gastrin is synthesized and stored in G cells in the antral mucosa of the stomach, duodenum and jejunum. Gastrin release into the blood is controlled in the following ways.

1. During the cephalic phase of digestion, cholinergic impulses in vagal nerves release gastrin. Hypoglycemia, a blood glucose concentration of 45 mg per 100 ml or less, acting through the hypothalamus, causes vagally mediated

release of gastrin. This fact is the basis for a test for the completeness of vagotomy. If the stomach fails to secrete acid and if the serum gastrin does not rise when adequate hypoglycemia is induced by insulin administration, vagotomy is presumed to be complete. However, an increase in acid secretion and a rise in serum gastrin during hypoglycemia is not proof of incompleteness, for gastrin release is stimulated by epinephrine which itself is liberated during profound hypoglycemia.

2. Secretagogues in contact with the antral mucosa release gastrin. The most potent are the amino acids, with phenylalanine at their head and valine trailing last. The L-isomers are more potent than the D-isomers. Polypeptide products of protein digestion are powerful stimulants, but native proteins are not. Glucose and fat have minor effects.

3. A rise in plasma ionized calcium releases gastrin, and consequently milk, the staple of traditional anti-ulcer regimens, stimulates acid secretion. So does coffee, not because it contains caffeine, but by action of its other constituents.

4. Aliphatic alcohols, of which ethanol is the most potent, release gastrin in the dog, a confirmed teetotaller, but not in man. Distention of the stomach stimulates acid secretion in man through local and vagal reflexes, but it liberates little if any gastrin. Secretin, when infused in pharmacological doses, inhibits gastrin release, and secretin liberated during digestion may participate in shutting off acid secretion.

Gastrin is liberated into the lumen of the stomach as well as into the blood, and gastrin in gastric contents, provided it is not destroyed by pepsin, stimulates acid secretion. The significance of these facts is unclear.

Chyme in the duodenum, particularly on account of its content of fat and protein digestion products, causes the gallbladder to contract, and in the 1920s the effect was demonstrated to be hormonally mediated. The responsible agent was called *cholecystokinin.* Much later, secretion of enzymes by the pancreas in response to irrigation of the duodenum with protein digestion products was found to be mediated by a hormone, and the agent was called *pancreozymin*. The two actions were eventually shown to be different properties of the same molecule, now called cholecystokinin-pancreozymin, or, more conveniently, CCK-PZ or CCK.

CCK is released into the circulation from cells in the duodenal and jejunal mucosa. The most potent stimuli are fat and protein digestion products.

A synthetic octapeptide, called OP-CCK, has the same active sequence of amino acids, and it is used experimentally and clinically as a substitute for CCK. The frog, *Hyla caerulea*, although ugly and venomous, bears within its skin a precious jewel, a decapeptide called *caerulein*, which, because it shares the C-terminal structure of CCK, is a powerful and clinically useful stimulant of gallbladder contraction.

Secretin is a polypeptide containing 27 amino acids. It has no subunit with biological activity; the entire molecule is required. Fourteen of its amino acids occupy the same position, counting from the N-terminus, as they do in glucagon, and secretin has metabolic effects like those of glucagon.

Secretin is released into the circulation from cells in the duodenal and jejunal mucosa by acid in contact with the mucosa. The amount of secretin released is proportional to the amount of acid.

The action of secretin in stimulating secretion of bicarbonate-containing juice from the pancreas and liver is strongly augmented by CCK.

During the course of early embryological development, cells that will be parent cells of hormone-secreting ones in the viscera migrate from the neural crest to organs of the digestive tract, and there they differentiate into cells with specific endocrine functions. Some become cells in the Isles of Langerhans which secrete insulin or glucagon. Others become gastrin- or secretin-secreting cells. Because the cells have a common origin and common histochemical characteristics, they are said to belong to the A.P.U.D series. This ridiculous name comes from *Amine Precursor Uptake* and *Decarboxylation*, and it is too firmly embedded in the literature to be quickly discarded.

Errors in differentiation and control sometimes occur, and a discrete tumor, called a gastrinoma, may synthesize and secrete gastrins at a very high rate. So may tumor cells diffusely spread through the pancreas and other digestive organs. Infusion of secretin stimulates release of gastrin from a gastrinoma. This is in contrast with secretin's inhibition of gastrin release from normal G cells, and the difference is the basis of a clinical test for the presence of a gastrinoma.

Stimulated by gastrin from a gastrinoma, the oxyntic mucosa secretes acid continuously in large amounts. This acid often overwhelms the capacity of the intestine to neutralize it, and the pH of intestinal contents may be very low as far as the midjejunum. Duodenal and jejunal mucosa become severely ulcerated. Pancreatic enzymes are denatured, and maldigestion may result in steatorrhea and azotorrhea. When the aberrant cells form a discrete tumor, they may be surgically removed, but when they are diffusely distributed or have metastasized, complete gastrectomy is the only successful treatment. The condition is called the Zollinger-Ellison syndrome after the surgeons who described it.

Gastrin is a trophic (or tropic) hormone for the oxyntic cells, and in persons who continuously have high concentrations of gastrin, there is hypertrophy and hyperplasia of the oxyntic mucosa. Capacity to secrete acid is increased.

Other cells of the A.P.U.D. series also form tumors which secrete one or several hormones. Of these, vasoactive intestinal peptide (VIP) stimulates the intestinal mucosa to secrete enormous amounts of bicarbonate-containing fluid, and the resulting diarrheal state is incorrectly called *pancreatic cholera*.

Hyperplasia of several peptide-secreting endocrine glands sometimes occurs, and a patient has *multiple endocrine adenomatosis*. In one combination, hyperplasia of parathyroid glands results in excessive secretion of parathormone, and the consequently elevated plasma calcium concentration stimulates secretion of gastrin. Such a patient may have a peptic ulcer.

Because peptide-secreting cells of the gut are nerve cells by origin, it is not surprising that gastrointestinal peptide hor-

TABLE 5-3.—SOME CANDIDATE HORMONES
OF THE GUT

NAME	SOURCE AND FUNCTION
Bombesin	In amphibian skin and antral and duodenal mucosa, probably in mammalian intestinal mucosa; stimulates gastrin release
Bulbogastrone	In mucosa of duodenal bulb; released by acid, inhibits gastric secretion
Chymodenin	In duodenal mucosa; stimulates pancreatic secretion of chymotrypsinogen
Duocrinin and enterocrinin	In duodenal mucosa; stimulate intestinal secretion
Enteroglucagon	In duodenal mucosa; mimics glucagon
Gastrin inhibitory polypeptide (GIP)	In duodenal mucosa; released by fat and carbohydrate; inhibits gastric motility, augments insulin release by glucose
Gastrone	In mucous fraction of gastric juice; inhibits gastric secretion
Incretin	In duodenal mucosa; releases insulin
Motilin	In duodenal mucosa; stimulates gastric motility, but inhibits gastric emptying
Pancreatic peptide	In pancreatic islets; stimulates acid secretion, but inhibits gastrin-stimulated secretion, inhibits basal pancreatic secretion
Somatostatin	In intestinal mucosa and nerves as well as in hypothalamus; multiple inhibitory actions
Urogastrone	In human male urine; identical with epithelial growth factor; inhibits gastric secretion
Vasoactive intestinal peptide (VIP)	In entire gut mucosa and nerves of intrinsic plexuses and of gallbladder; vasodilator, inhibits acid and pepsin secretion, mimics glucagon and may be neurotransmitter
Villikinin	In duodenal mucosa; stimulates contraction of intestinal villi

mones (gastrin and VIP, for example) are found in the brain and that VIP and somatostatin are found in nerve cells of the gut.

Many peptides have been isolated from gastrointestinal tissue, and the structures of many have been determined. When injected into experimental animals, they evoke a large number of physiological responses, and they are often assumed to be hormones mediating those responses. So far, most are only "candidate hormones," waiting to satisfy all the criteria for admission to the lodge. The criteria include proof that a candidate hormone has qualitative and quantitative properties adequate to explain its participation in a known endocrine system. A partial list of candidate hormones and their properties is given in Table 5–3.

SUGGESTED READINGS

Dockray, G. J.: Molecular evolution of gut hormones: Application of comparative studies on regulation of digestion, Gastroenterology 72:344, 1977.

Grossman, M. I., et al.: Candidate hormones of the gut, Gastroenterology 67:730, 1974.

Johnson, L. R.: Gastrointestinal hormones and their functions, Ann. Rev. Physiol. 39:135, 1977.

Johnson, L. R.: The trophic action of gastrointestinal hormones, Gastroenterology 70:278, 1976.

Johnson, L. R. (ed.): Gastrointestinal hormones: physiological implications, Fed. Proc. 36:1929, 1977.

Pearse, A. G. E., Polak, J. M., and Bloom, S. R.: The newer gut hormones. Cellular sources, physiology, pathology, and clinical aspects, Gastroenterology 72:746, 1977.

Said, S., and Rosenberg, R. N.: Vasoactive intestinal polypeptide: Abundant immunoreactivity in neural cell lines and normal nervous tissue, Science 192:908, 1977.

Walsh, J. H., and Grossman, M. I.: Gastrin, N. Engl. J. Med. 292: 1324, 1377, 1975.

6. GASTRIC SECRETION

The two main divisions of the gastric mucosa are the oxyntic glandular mucosa and the pyloric glandular mucosa (Fig 6 – 1). The first, which secretes acid, covers the body of the stomach, and the second covers the antrum. The distribution of the two types of mucosa, however, does not exactly correspond to the two divisions of the stomach based on the nature of the muscular layer.

The surface of the oxyntic glandular area is covered with a layer of tall, columnar surface epithelial cells that contain and secrete mucus (Fig 6 – 2). The surface is thickly studded with pits into which long tubular glands empty. At the junction of pit and gland are the neck chief cells. These, too, contain and secrete mucus, but their most important function is to serve as parent cells for the replacement of other cells of the gastric mucosa. The chief cells forming the walls of the tubules are the ones which synthesize and secrete the enzyme precursor, pepsinogen. Along the outside wall of the glands are the cells which secrete hydrochloric acid. Because they are on the wall, they are called parietal cells (Latin, *paries* = wall), and because they secrete acid they are called oxyntic cells (Greek, *oxy* = sharp). The latter name will be used here. In man the oxyntic cells also secrete the intrinsic factor, necessary for absorption of vitamin B_{12} in the terminal ileum. Other cells of the oxyntic glandular area contain large amounts of histamine, 5-hydroxytryptamine and heparin. Liberation of these compounds is important in pathological conditions.

The surface of the pyloric glandular mucosa is also covered with surface epithelial cells that contain and secrete mucus. Secretion of mucus is copious during digestion of a meal, and mucus appears to be an important lubricant of the pyloric glandular mucosa. Cells of the pyloric glands secrete a small amount of a neutral fluid roughly similar to an ultrafiltrate of

45

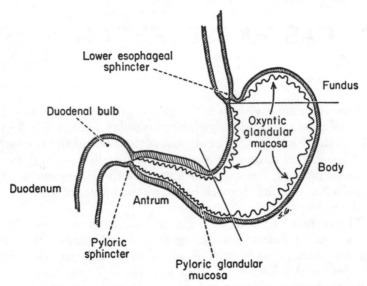

Fig 6–1.—The parts of the stomach and duodenum. The border between the pyloric glandular mucosa and the oxyntic glandular mucosa does not exactly coincide with the border between antrum and body.

plasma. They also secrete a small amount of pepsinogen, which is chemically different from that secreted by the oxyntic glandular mucosa.

The oxyntic glandular mucosa secretes a juice containing H^+, Cl^-, Na^+ and K^+. As the rate of secretion increases, the concentration of H^+ rises, and the concentration of Na^+ falls. At the highest rate of secretion, the fluid collected from the stomach is a nearly isotonic solution containing HCl at approximately 145 mN (millinormal) and KCl at approximately 10 mN (Fig 6–3).

Samples of gastric juice taken from the stomach may have been diluted and partially neutralized by swallowed food and saliva or by regurgitated duodenal contents. When these factors are eliminated, the composition of gastric juice varies with the rate of secretion. There are three explanations for this.

1. Most physiologists believe that the acid fluid secreted by

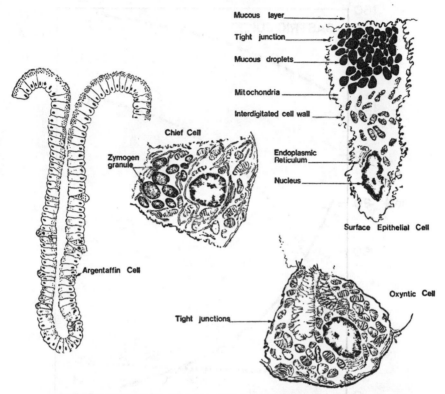

Fig 6–2.—A gland of the oxyntic portion of the gastric mucosa. Surface epithelial cells cover the gastric pits and the ridges between them. Surface cells contain and secrete mucus which covers the surface of the mucosa, and the cells are attached to one another by highly impermeable tight junctions. The chief cells, which secrete pepsinogens, line the lower half of the gland. The oxyntic, or parietal cells, which secrete acid, occupy a peripheral position between the chief cells. They are packed with mitochondria, and when they are secreting, numerous intracellular canaliculi lead to the lumen of the gland. Argentaffin cells are scattered singly between the chief cells.

the oxyntic cells is constant in composition. If this is true, either of the other two explanations will account for the variability of the fluid collected. There has never been any direct experimental proof, however, of the constancy of composition of the fluid as it emerges from the secreting

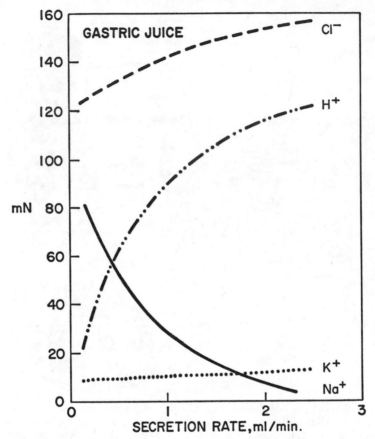

Fig 6–3.—The relationship between concentration of electrolytes in the gastric juice of a normal young man and the rate of secretion. Secretion was stimulated by intravenous infusion of histamine at various constant rates, and the juice to be analyzed was collected after steady rate of secretion had been reached. (Adapted from Nordgren, B.: Acta Physiol. Scandinav., Suppl. 202, 1963.)

cells. Approximate constancy is probable, but absolute constancy remains to be demonstrated.

2. Other cells of the mucosa, the surface epithelial cells, the neck chief cells, the chief cells, secrete small volumes of fluid the electrolyte composition of which is thought, on evidence that is not completely convincing, to be similar to

an ultrafiltrate of plasma. As such, it contains bicarbonate at about 24 mN, and bicarbonate neutralizes an equivalent amount of acid. The fluid also contains Na^+ at about 145 mN. This fluid is supposed, again on not completely convincing evidence, to be secreted at a constant rate when acid secretion is stimulated. If a fluid having the composition of an ultrafiltrate of plasma and secreted at a constant rate is mixed with acid juice secreted at a variable rate, the resulting relationship between composition of the juice collected from the stomach and its rate of secretion should be similar to that actually found (Fig 6–3).

3. The surface of the gastric mucosa is only very slightly permeable to H^+, and as gastric juice flows over the surface of the stomach, acid slowly leaves the juice by diffusing back into the mucosa. The mucosa is also only very slightly permeable to Na^+ and Cl^- contained in its interstitial fluid. As acid gastric juice flows over the surface of the stomach, Na^+ slowly diffuses into gastric juice in exchange for H^+, and an additional small amount of Na^+ and Cl^- diffuse together from interstitial fluid into gastric juice. These processes of diffusion may also explain the relationship between the rate of secretion of gastric juice and its composition.

Explanations 2 and 3 are not mutually exclusive.

Gastric mucosal cells contain an electrogenic chloride pump. When there is little or no stimulus for secretion, this pump is minimally active, but because it continues to secrete Cl^- into the lumen at a low rate, the luminal surface of the oxyntic glandular mucosa is negative with respect to the serosal surface or the blood by about 40 to 60 mV.

When the oxyntic cells are stimulated to secrete acid, the electrogenic H^+ pump becomes active, and the Cl^- pump increases its rate of pumping so that H^+ and Cl^- are secreted together. Then the potential difference across the mucosa falls slightly. (In experimental conditions in vitro, the gastric mucosa can be made to secrete H^+ without any accompanying Cl^-, and in that case the luminal surface of the mucosa becomes positive with respect to the serosal surface.)

An oxyntic cell secretes H^+ into intracellular canaliculi, and

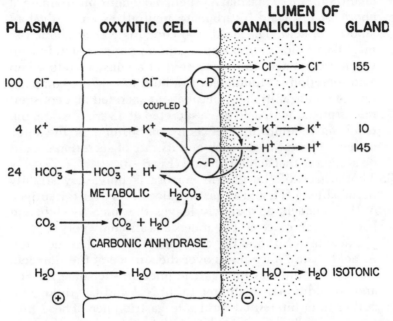

Fig 6-4.—Processes involved in secretion of acid by an oxyntic cell. When the cell is not stimulated, the electrogenic chloride pump is minimally active, and a potential difference is established across the cell. When the cell is stimulated, the electrogenic hydrogen ion pump secretes H^+, probably by exchanging H^+ for K^+, and the potential difference falls slightly. The chloride pump, which is somehow coupled with the hydrogen ion pump, increases its activity, and the two ions are secreted together. Carbonic acid, derived from metabolic and blood-borne carbon dioxide, furnishes hydrogen ions to replace those secreted and bicarbonate ions to replace chloride taken from the plasma. Flow of water through the cell makes the secreted juice nearly isotonic. The numbers in the margins give the approximate concentrations of the ions in plasma and gastric juice in milliequivalents per liter.

juice formed in the canaliculi flows into the lumen of the gastric glands and thence to the surface of the mucosa. The canalicular membrane contains a K^+-ATPase, and when K^+ is on the luminal side of the membrane, K^+ is transported into the cell in exchange for H^+ which is secreted into the canaliculus

(Fig 6–4). It is likely that the K^+ which is absorbed as H^+ is secreted is derived from the same oxyntic cell. Some K^+ escapes into the lumen of the gland.

ATP, which drives the exchange of H^+ for K^+, is made by the numerous mitochondria which pack the cell. Neutrality of the interior of the cell is maintained by replacing the H^+ secreted with H^+ derived from carbonic acid. Carbon dioxide from blood and from cellular metabolism is hydrated to carbonic acid, and hydration is catalyzed by carbonic anhydrase present in the same membrane that secretes H^+. Carbonic acid ionizes, giving an H^+ to replace the one secreted, and an HCO_3^- which passes into the plasma to replace the Cl^- accompanying H^+ into the juice. Water flows from plasma to juice as the ions are secreted, and the juice is very nearly isotonic with plasma in gastric capillaries.

In the process of secretion of acid juice, the metabolic machinery of the oxyntic cells manufactures new ions, H^+ and HCO_3^-, which are osmotically active. Consequently, as secretion proceeds, the total osmotic pressure of the plasma rises.

When gastric juice is lost, either by vomiting or by drainage, metabolic alkalosis results, because for each H^+ lost in the juice, one HCO_3^- has been retained in the blood. Loss of gastric juice also causes dehydration and shrinkage of extracellular fluid volume, and loss of K^+ may result in a negative potassium balance with all its consequences.

Gastric secretion is stimulated in three ways.

1. *By gastrins*. Endogenous gastrins liberated from antral and intestinal mucosa and exogenous gastrins and pentagastrin are powerful stimulants of acid secretion. Gastrins also stimulate secretion of pepsinogens.
2. *By acetylcholine*. Acetylcholine liberated by nerve endings of the intrinsic plexuses stimulate secretion of acid, pepsinogens and mucus. Long-lived cholinomimetic drugs, such as bethanechol, are also effective, and atropine and atropine-like drugs inhibit secretion. Acetylcholine and its congeners act synergistically with gastrins and pentagastrin, and a background of cholinergic activity enhances the

effect of circulating gastrins. Consequently, atropine inhibits gastrin-stimulated secretion.

3. *By histamine.* Histamine, which is itself a powerful stimulant of acid secretion and a weak stimulant of pepsinogen secretion, is stored in large quantities within cells of the oxyntic mucosa, and the mucosa is capable of transforming histamine into methylated derivatives which are even more powerful stimulants. Histamine occurring naturally within the mucosa can act in one of two ways. (1) Histamine may be the *final common mediator,* and other stimuli, gastrins and acetylcholine, may act only by liberating histamine. This is probably not the case. (2) Histamine, liberated continuously during the basal state and perhaps liberated in increasing amounts when natural stimuli act, can sensitize the oxyntic cells to other stimuli. In this case, the efficacy of other stimuli depends upon the concurrent efficacy of histamine. If histamine is prevented from acting on the secretory cells, the other stimuli are ineffective as well. Therefore, antihistaminic drugs would inhibit secretion stimulated by acetylcholine and by gastrins.

The general run of antihistaminic drugs blocks the action of histamine on many cells, and they are said to compete with histamine for H_1 receptors on those cells. Those drugs do not block the action of histamine upon oxyntic cells, and consequently histamine is said to combine with different receptors, called H_2 receptors, on oxyntic cells. A new class of histamine antagonists, of which cimetidine is an example, are competitive inhibitors of histamine-stimulated acid secretion, and they are called H_2 blockers. These drugs not only inhibit histamine-stimulated secretion, they are powerful inhibitors of gastrin- and acetylcholine-stimulated secretion as well. This fact is believed to support the opinion that the natural effect of histamine is to permit gastrins and acetylcholine to be effective stimuli.

Gastric secretion is naturally stimulated in the following ways (Fig 6-5).

1. *Cephalic phase.* Stimuli acting in the head, among them

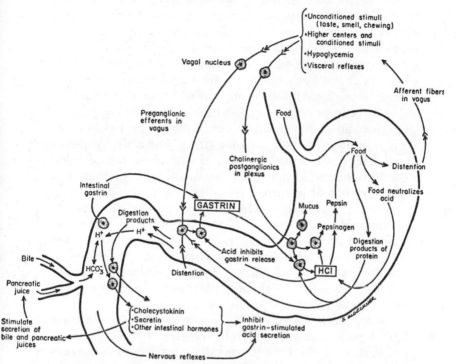

Fig 6-5.—The major factors controlling gastric secretion.

smells, tastes, chewing and swallowing food, cause impulses to flow along vagus nerves to the stomach. In the body of the stomach, these impulses cause the release of acetylcholine near secretory cells, stimulating secretion of acid, pepsinogens and mucus. In the antrum, the impulses promote liberation of gastrin, provided the pyloric glandular mucosa is not bathed by an acid solution. When it acts alone, the cephalic phase stimulates secretion at about half the maximal pentagastrin-stimulated rate. Under ordinary circumstances, the cephalic phase is covered by the subsequent gastric phase.

2. *Gastric phase.* Food in the stomach stimulates gastric secretion by three means. (1) Distention of the stomach stim-

ulates mechanoreceptors, the afferent fibers of which travel centrally in the vagus nerve. Resulting efferent discharges, also in the vagus nerve, stimulate gastric secretion. (2) Distention of the stomach stimulates mechanoreceptors, the afferent impulses of which increase activity in the intrinsic plexuses of the body of the stomach and stimulate gastric secretion of a local reflex. In man, distention of the stomach releases little or no gastrin. (3) Protein digestion products, peptides and amino acids, in contact with the pyloric mucosa stimulate the release of gastrin, probably by direct contact with the apical surface of gastrin-secreting cells. Gastrin is liberated into both the blood and the lumen of the stomach, and in turn it stimulates acid secretion. Gastrin release is inhibited when the contents of the antrum become acid.

During digestion of a protein-containing meal, the rate of acid secretion approaches the maximal response to pentagastrin. Fats and carbohydrates do not stimulate acid secretion, and, in man, an ethanol solution has no more effect than an equal volume of Adam's ale.

3. *Intestinal phase.* Chyme in the upper small intestine stimulates a gradual increase in acid secretion which reaches a peak of about 30% of the pentagastrin maximum in the second hour. At least one mediator is big gastrin, but gastrin release does not account for all the intestinal phase. Other hormones, as yet incompletely identified, and perhaps absorbed amino acids, may be additional stimuli.

Gastric secretion is inhibited by the following means:

1. *Exogenous drugs.* Atropine and antihistaminics acting on H_2 receptors inhibit acid secretion. Prostaglandins of the 16,16-dimethyl-E_2 type, in minute amounts, are powerful inhibitors. Inhibitors of carbonic anhydrase, when given in massive doses, inhibit acid secretion, but this is unimportant.

2. *Reflexes.* Acid, fat digestion products and solutions of high osmotic pressure in the duodenum and upper jejunum inhibit gastric secretion, in part through nervous reflexes that

are poorly understood. There are noncholinergic, nonadrenergic fibers in the vagus nerve which inhibit gastric secretion when stimulated, and sympathetic nerves and intrinsic plexuses may also participate.

3. *Hormones.* CCK-PZ released from the intestinal mucosa competitively inhibits gastrin-stimulated secretion of acid. The basis for this is competition for receptor sites on the secreting cells. In order to stimulate copious secretion of acid, gastrin must combine with a receptor on the cell. CCK-PZ has the same amino acid sequence in its active group, and it too combines with the same receptor. CCK-PZ, however, is only a weak stimulant of acid secretion. By occupying the receptor site, CCK-PZ denies the site to gastrin, a much stronger stimulant of acid secretion. Secretin also inhibits gastrin-stimulated acid secretion. Because the structure of secretin has no active group in common with that of gastrin, one would expect the inhibition to be noncompetitive. It is so in the dog, but in man it is competitive with gastrin. Somatostatin also competitively inhibits gastrin-stimulated acid secretion, and other as yet incompletely characterized hormones from the duodenal mucosa and elsewhere may also be physiological inhibitors.

Before the isolation and identification of CCK-PZ and secretin, crude extracts of the duodenal mucosa were found to inhibit gastric secretion and motility. The supposed active principle was called *enterogastrone.* At least part of the effects of enterogastrone can be attributed to the extracts' content of CCK-PZ and secretin, but there are other compounds in such extracts that also inhibit gastric secretion and motility. Their physiological role is at present unknown.

The rate of acid secretion rises to a maximum in the second half hour after the beginning of a meal, and it declines slowly over the next several hours. The total amount of acid secreted in response to a meal is directly proportional to the protein content of the meal (Fig 6–6). The reason is that protein is the best buffer contained in food. Dilution and neutralization of residual acid in gastric contents at the beginning of a meal allow gastrin to be released. Gastrin stimulates secretion of acid,

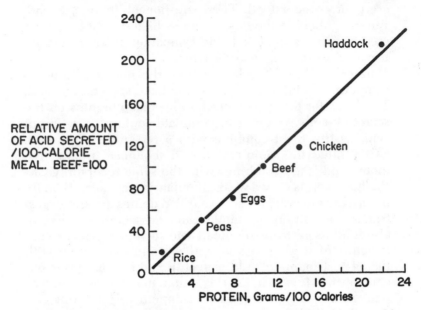

Fig 6–6.—The relationship between the amount of acid secreted by the stomach during digestion of a meal and the protein content of the food. The amount of acid secreted in response to a 100-calorie meal of beef is put at 100 and is plotted against the protein content of the beef used, 10.7 gm per 100 calories. The relative amounts of acid secreted in response to 100-calorie meals of the foods named are plotted against the protein content of the foods. Points obtained with 24 other foods having a wide range of protein content fall close to the line drawn in the figure. (Adapted from Saint-Hilaire, S., et al.: Gastroenterology 39:1, 1960.)

and acid titrates the buffers of the food. Eventually, when the buffers have been titrated to a low pH, acid in contact with the pyloric glandular mucosa inhibits further release of gastrin, and a major stimulant of acid secretion is withdrawn. Any other buffer or neutralizing agent added to food has the same effect as protein in governing the amount of acid secreted.

When the stomach has emptied and stimuli for acid secretion have been withdrawn, the stomach is left with a small

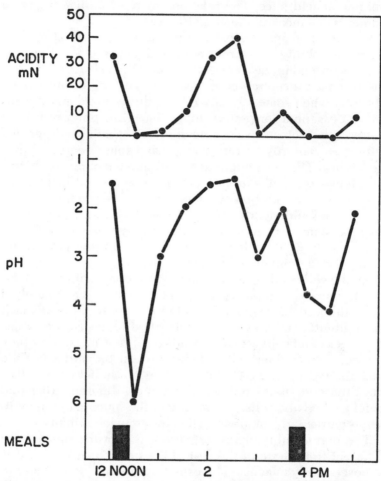

Fig 6–7.—The acidity of contents removed from the human stomach by tube expressed as millinormal acid and as pH. At 12 noon, there was only a small amount of acid fluid in the stomach. After a luncheon of fish, potato, carrot, fruit, custard and tea, the acidity of the contents of the stomach was very low. The acidity gradually climbed as acid was secreted in response to the meal and as the stomach emptied. Acidity fell once more at 4 P.M., when bread and butter were eaten with tea. (Adapted from James, A. H., and Pickering, G. W.: Clin. Sci. 8:181, 1949.)

volume of acid juice. The relation between acidity of gastric contents and meals is shown in Figure 6–7.

In some persons, the oxyntic glandular mucosa atrophies, losing its ability to secrete acid and pepsinogen. Because there is no acid in the stomach to inhibit the release of gastrin, the plasma concentration of gastrin is very high, but it falls abruptly when some 100 mN HCl is drunk. Acid and pepsin are not essential for protein digestion. Absorption of iron is better if the stomach is capable of secreting acid, and persons with gastric mucosal atrophy may have some degree of iron deficiency. The most important consequence of mucosal atrophy, however, results from its failure to secrete intrinsic factor necessary for absorption of vitamin B_{12} in the terminal ileum, and persons with deficiency of intrinsic factor have pernicious anemia. Many persons with gastric atrophy have antibodies to oxyntic cells in their blood, and gastric atrophy may be the result of an autoimmune process.

All the cells of the gastric mucosa turn over rapidly, and the whole mucosa is replaced in about 3 days. Approximately a half million cells desquamate into the lumen of the stomach each minute, and they can be identified in gastric washings. The neck chief cells are the parents of all new cells. After they divide, the daughter cells migrate toward the surface of the mucosa and differentiate into surface epithelial cells. Other daughter cells migrate down the glands, differentiating into chief and oxyntic cells. The ability of the mucosa to renew itself is particularly important when the stomach is injured.

The permeability characteristics of the gastric mucosa are different from those of the rest of the gut. For instance, it is almost entirely unaffected by osmotic gradients. Pure water in the stomach is not absorbed, and if the contents of the stomach are hypertonic, little water moves from blood into the lumen. This is in contrast with the duodenum where hypotonic or hypertonic solutions are quickly brought to isotonicity.

The gastric mucosa is normally only very slightly permeable to the acid which it secretes. This property, called the *gastric mucosal barrier*, accounts for the stomach's ability to contain acid without injuring itself. Many substances, however, break

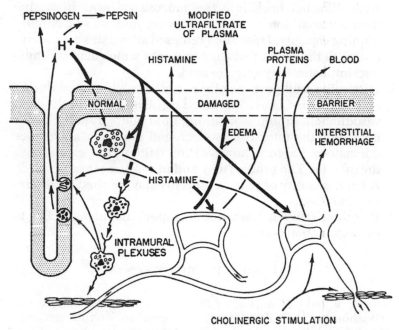

Fig 6-8. — Pathophysiological consequences of the back-diffusion of acid through the broken gastric mucosal barrier.

the gastric mucosal barrier. The most important of these are salicylates: aspirin and salicylic acid in acid but not in neutral solution. Other important compounds which break the barrier are ethanol and regurgitated bile acids and lysolecithin of duodenal contents. In some otherwise normal persons, the barrier may be, for unknown reasons, weak, and it frequently breaks in patients with severe injuries or infections. When the barrier is broken, acid can diffuse back into the mucosa with serious pathophysiological consequences (Fig 6-8).

1. Acid diffusing back into the mucosa stimulates motility of the stomach by acting on the intrinsic plexuses. Strong contractions by an indurated stomach may give rise to pain.
2. Acid diffusing back into the mucosa stimulates secretion of pepsinogen.

3. Acid diffusing back into the mucosa releases histamine from mucosal stores and increases the mucosa's histamine-forming capacity. Histamine released stimulates acid secretion with the result that a mucosa into which acid is diffusing simultaneously secretes acid.

4. Histamine, and perhaps other substances liberated during mucosal injury, cause increased capillary permeability and vasodilation. Protein-containing fluid pours from the capillaries into the interstitial spaces, and the mucosa becomes edematous. Large volumes of interstitial fluid exude from the mucosa. The process may end in capillary stasis.

5. A large quantity of plasma proteins may be shed by the injured mucosa.

6. Bleeding ranging from minor superficial hemorrhage to exsanguination occurs.

SUGGESTED READINGS

Code, C. F. (ed.): *Handbook of Physiology: The Alimentary Canal* (Washington, D.C.: American Physiological Society, 1968), Chaps. 41–51, 77, 138.

Longstreth, G. F., Go, V. L. W., and Malagelada, J-R.: Postprandial gastric, pancreatic, and biliary response to histamine H_2-receptor antagonists in active duodenal ulcer, Gastroenterology 72:9, 1977.

Pounder, R. E., Williams, J. G., Russell, R. C. G., Milton-Thompson, G. J., and Misiewicz, J. J.: Inhibition of food-stimulated gastric acid secretion by cimetidine, Gut 17:161, 1976.

Richardson, C. T., Walsh, J. H., Cooper, K. A., Feldman, M., and Fordtran, J. S.: Studies on the role of cephalic-vagal stimulation on the acid secretory response to eating in normal human subjects, J. Clin. Invest. 60:435, 1977.

Richardson, C. T., Walsh, J. H., Hicks, M. I., and Fordtran, J. S.: Studies on the mechanisms of food-stimulated gastric acid secretion in normal human subjects, J. Clin. Invest. 58:623, 1976.

Robert, A., Schultz, J. R., Nezamis, J. E., and Lancaster, C.: Gastric antisecretory and antiulcer properties of PGE_2, 15-methyl PGE_2, and 16,16-dimethyl PGE_2. Intravenous, oral and intrajejunal administration, Gastroenterology 70:359, 1976.

Walsh, J. H., Richardson, C. T., and Fordtran, J. S.: pH dependence of acid secretion and gastrin release in normal and ulcer subjects, J. Clin. Invest. 55:462, 1975.

7. GASTRIC MOTILITY

With each swallow as a meal is eaten, there is a slight receptive relaxation of the body of the stomach so that the stomach accommodates itself to the meal with little rise in intragastric pressure. Receptive relaxation is a vagally mediated reflex. Throughout the digestion of a meal, only feeble mixing movements, if any, occur in the body of the stomach. If the contents of the body of the stomach are foods having a high viscosity — such as meatballs or mashed potatoes — the contents are poorly mixed with acid secretion, and they remain neutral. Peristaltic waves move over the antrum to the pyloric sphincter, mixing food with digestive juices and slowly propelling the mixture into the duodenum. As the volume of the contents decreases, the wall of the body gently contracts.

When the contents of the body are not mixed, digestion of protein is slight or absent. Acid juice secreted by the oxyntic glandular mucosa flows around the mass of food to the antrum where it is thoroughly mixed with small amounts of food. Digestion of starch by ptyalin continues in the body of the stomach until it is stopped by slowly rising acidity, whereas digestion of protein begins in the antrum.

Peristalsis is governed by a wave of partial depolarization which begins in a group of pacemaker cells in the longitudinal muscle layer high on the greater curvature of the stomach. The wave sweeps over the longitudinal layer toward the pylorus, and it almost completely dies out at the pyloric sphincter. This BER, as it moves over the longitudinal muscle, may or may not be accompanied by contraction of the underlying circular muscle. Vagal fibers fan out over the antrum, and when they are sequentially active from above downward, the threshold of the circular muscle fibers is lowered. Then the circular muscle contracts in step with the BER, and a peristaltic wave moves over the antrum. If the circular muscle does not contract, the

BER sweeps on anyway with no visible sign of its presence. The frequency with which the BER originates is close to 3 per minute, so the rhythm of peristalsis in the stomach is the same. The velocity of the BER over body and antrum is so slow, about 1 cm per second, that two or three successive peristaltic waves may be seen at one time. The wave accelerates as it reaches the terminal antral segment, with the result that the terminal antrum and the pyloric sphincter contract almost simultaneously.

If contraction of the circular muscle does accompany the BER, the contraction may be weak or strong. The BER is a wave of partial depolarization, and consequently it is a current sink. Current flows from the surface of the circular muscle fibers to the sink, and this electrotonic flow of current tends to depolarize the circular muscle. Whether or not the circular muscle fibers are depolarized sufficiently to bring them to threshold so that they have action potentials and increase in tension depends upon the excitability of the circular muscle fibers at the time. If there is a lot of vagal excitatory activity, threshold will be low. Many circular muscle fibers will reach threshold, spike and contract. The wave of peristalsis will be strong. If there is little vagal activity or if there is sympathetic inhibitory activity, threshold will be high. Few or no circular muscle fibers will reach threshold, and peristaltic contractions will be weak or absent.

The body of the stomach is a hopper, and its contents are gradually fed into the antrum (Fig 7 – 1). No matter how strong an antral peristaltic wave may be, its ring of contraction is never closed. As a wave advances, it mixes food and digestive juices within the antrum, and the mixture escapes backward through the open ring, only to be tumbled again by the next peristaltic wave.

At rest, the muscle of the pyloric sphincter is either relaxed or only very slightly contracted. The pyloric canal is closed and empty. There is a zone of pressure within it which is a bit higher than the pressure in the antrum or duodenum, and therefore no chyme flows in either direction through the canal. As the peristaltic wave advances in the antrum, the viscos-

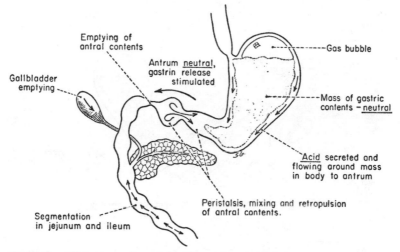

Fig 7–1.—The stomach and duodenum at the beginning of the digestion of a meal.

ity of the chyme it propels causes the pressure in front of the
canal to rise, and a small amount of chyme may pass through
the canal into the duodenum (Fig 7–2). Gentle contraction
of the body of the stomach helps to increase the pressure gradient between stomach and duodenum. How much chyme
passes through at any time depends upon the pressure gradient between antrum and duodenum. As the peristaltic wave
reaches the terminal antral segment, the pyloric sphincter
contracts and abruptly cuts off passage of chyme into the duodenum. The sphincter then relaxes to its resting pressure and
remains empty until the next peristaltic wave comes along.

Factors that influence the rate of gastric emptying do so
chiefly by affecting the pressure gradient between stomach
and duodenum. Pressure in the stomach is determined by tension in its wall. Other things being equal, the greater the volume of gastric contents, the greater is the tension. Consequently, for a liquid meal, the rate of emptying is greater the
larger the volume of gastric contents, and the rate of emptying
is greater at the beginning of the process than at the end (Fig

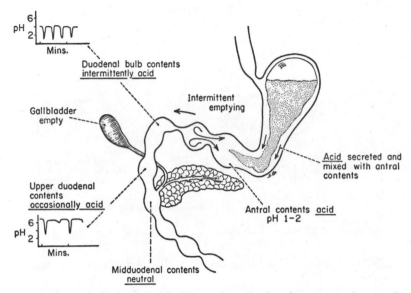

Fig 7–2.—Stomach and duodenum late in the digestion of a meal. The contents of the stomach have been titrated to pH 1–2 by secreted acid. The pH of the contents of the duodenal bulb falls to 2 when gastric contents pass through the pyloric sphincter, three times a minute. In the upper duodenum the pH falls occasionally, and by midduodenum, intestinal contents are neutral.

7–3). For this reason, the concept of "emptying time," the time the last trace of a meal leaves the stomach, has little meaning.

Solids or semisolids are emptied much more slowly than liquids (Fig 7–4). A solid piece of food is denied entry into the pyloric canal; it is pushed backward and forward in the antrum until it is sufficiently eroded to be emptied along with liquid. One quarter of a meal of cooked chicken livers chopped into 1 cm cubes leaves the stomach in each of 4 hours. Although an occasionally swallowed solid object may be recovered in the stool, safety pins or razor blades, swallowed with suicidal intent or in pica, may accumulate in large quantities in the stomach.

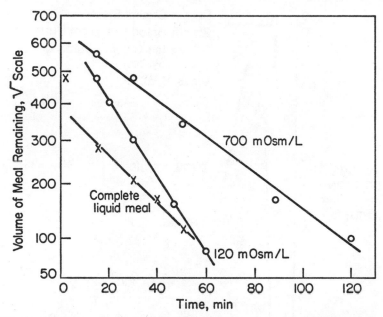

Fig 7–3.—The volume of gastric contents of a normal man plotted against time, showing that the rate of emptying is a linear function of the square root of the volume remaining in the stomach. The reason for the square root relation may be that the circumferential tension in the wall of a cylinder is proportional to the square root of its volume. The osmotic pressure of the hypertonic and hypotonic meals was adjusted by adding sucrose to a mixture of citrus pectin and water. The complete liquid meal was a milk product to which sugar had been added. (Adapted from Hunt, J. N., and Spurrell, W. R.: J. Physiol. 113:157, 1951; Hunt, J. N.: Gastroenterology 45:149, 1963; and Hopkins, A.: J. Physiol. 182:144, 1966.)

The rate of gastric emptying is governed by the ability of the duodenum to deal with chyme delivered to it. The most important qualities of chyme are its acidity, its osmotic pressure, its fat and protein content and its caloric density. Gastric emptying is slowed until chyme in the duodenum is neutralized and made isotonic. Fat digestion begins in the duodenum, and free fatty acids and 2-monoglycerides resulting from hydrolysis of triglycerides are powerful inhibitors of gastric

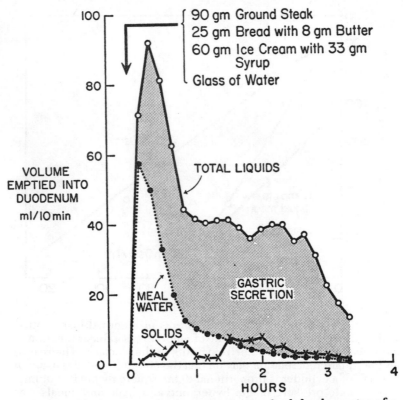

Fig 7–4.—Average rate of gastric emptying of solids, the water of a meal and the accompanying secretion. The data are the average found in 7 normal human subjects. (Adapted from Malagelada, J-R.: Gastroenterology 72:1264, 1977.)

emptying. So are some amino acids liberated during protein digestion. Slowing of gastric emptying is accomplished by reflexes and hormones originating in the duodenum and jejunum; they relax the wall of the stomach and diminish the strength of peristaltic contractions. In some unknown manner the upper small intestine also monitors the rate calories are delivered to it. A meal of high caloric density is delivered to the duodenum more slowly than one containing fewer calories per milliliter.

When the stomach is finally empty after a meal, peristaltic contractions die out. If fasting is prolonged for 10 or so hours, strong peristaltic waves again sweep over the antrum. These are the beginning of the interdigestive myoelectric complex described in Chapter 9. A fasting man may feel hunger pangs during a particularly vigorous series of peristaltic waves as afferent impulses are sent to the brain in vagal fibers.

SUGGESTED READINGS

Code, C. F. (ed.): *Handbook of Physiology: The Alimentary Canal* (Washington, D.C.: American Physiological Society, 1968), Chaps. 93–97, 104, 139.

Cannon, W. B.: *The Mechanical Factors of Digestion* (London: Edward Arnold, 1911).

Cooke, A. R.: Localization of receptors inhibiting gastric emptying in the gut, Gastroenterology 72:875, 1977.

Hunt, J. N., and Stubbs, D. F.: The volume and energy content of meals as determinants of gastric emptying, J. Physiol. 245:209, 1975.

MacGregor, I. L., Martin, P., and Meyer, J. H.: Gastric emptying of solid food in normal man and after subtotal gastrectomy and truncal vagotomy with pyloroplasty, Gastroenterology 72:206, 1977.

Malagelada, J-R.: Quantification of gastric solid-liquid discrimination during digestion of ordinary meals, Gastroenterology 72:1264, 1977.

Malagelada, J-R., Longstreth, G. F., Summerskill, W. H. J., and Go, V. L. W.: Measurement of gastric functions during digestion of ordinary solid meals in man, Gastroenterology 70:203, 1976.

Stubbs, D. F.: Models of gastric emptying, Gut 18:202, 1977.

8. PANCREATIC SECRETION

The external secretion of the pancreas contains two components.

1. One is an aqueous secretion whose most important characteristic is that it contains a high concentration of bicarbonate. In man, it is secreted at the rate of 200–800 ml a day. The liver secretes a similar fluid, in addition to the fluid secreted with bile acids, and secretion of this fluid by the liver is under the same control as that of the aqueous secretion of the pancreas. When the aqueous secretion of the pancreas is discussed, it is to be understood that the aqueous, bile acid-independent secretion of the liver is included.

2. The other is a solution of small volume containing all the enzymes and enzyme precursors synthesized and secreted by the acinar cells of the pancreas. This secretion is swept into the duodenum by concurrently secreted aqueous juice.

The aqueous component is an isotonic solution whose major cation is sodium. It also contains a small amount of potassium, calcium and magnesium. The anions are bicarbonate and chloride, and their sum equals the sum of the cations. As the rate of pancreatic secretion increases, the concentration and rate of output of bicarbonate rise, and the concentration of chloride falls (Fig 8–1). The function of this juice is to neutralize acid entering the duodenum from the stomach.

When stimulated by the hormone secretin, centroacinar cells within pancreatic acini secrete fluid containing more chloride than bicarbonate. Under the same stimulus, cells lining the immediate extralobular ducts secrete a greater volume of fluid containing a high concentration of bicarbonate and a low concentration of chloride (Fig 8–2). Cells lining the more distal ducts are permeable to anions, and bicarbonate in the

69

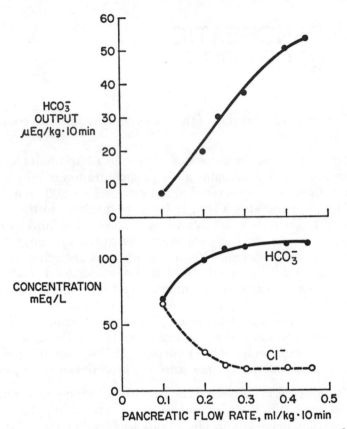

Fig 8–1.—*Top*, the relation between the bicarbonate output of the pancreas and the rate of secretion of pancreatic juice. Each point is the average of samples collected by endoscopic cannulation of the ampulla of Vater in five normal human subjects whose pancreatic secretion was stimulated by intravenous administration of secretin. *Bottom*, the relation between bicarbonate and chloride concentrations in the same samples of pancreatic juice. (Adapted from Domschke, S., et al.: Gastroenterology 73:478, 1977.)

fluid within the ducts exchanges for chloride from the extracellular fluid. Consequently, fluid entering the main ducts, where no secretion or exchange occurs, is a mixture of two secretions modified by diffusion. When fluid is secreted at a low rate, diffusion dominates, and pancreatic juice has a relatively low concentration of bicarbonate. When secretion is rapid, there is less time for diffusion, and the bicarbonate concentration of pancreatic juice is high.

Pancreatic secretion is stimulated by impulses arriving in cholinergic vagal fibers and by gastrin liberated from antral and duodenal mucosa. Consequently, there is a cephalic phase of pancreatic secretion which is about 25% of the response to a meal. Fat and protein digestion products in the duodenum also stimulate pancreatic secretion by means of a cholinergic reflex whose afferent and efferent limbs are both in the vagus.

The major stimuli for pancreatic secretion are the hormones secretin and CCK-PZ. Secretin is liberated from the duodenal

Fig 8–2.—Secretion by centroacinar cells and by cells of the extralobular ducts of the pancreas. Chloride concentrations given on the right were determined on fluid collected by micropuncture, and bicarbonate concentrations were inferred from the fact that the fluid is isotonic. These data are for the cat pancreas, but other species seem to be similar. (Adapted from Lightwood, R., and Reber, H. A.: Gastroenterology 72:61, 1977.)

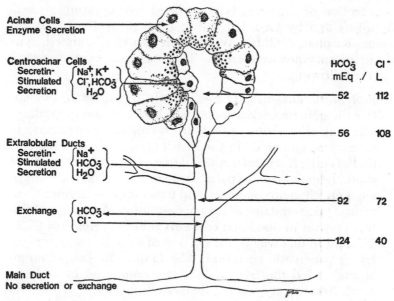

mucosa by acid, and the amount of secretin liberated is proportional to the amount of acid entering the duodenum. Secretin is a potent stimulant of pancreatic aqueous secretion, and in large doses it causes maximal secretion. In the normal course of digestion, not enough secretin is liberated to account for the amount of pancreatic juice secreted. The reason is that most of the acid arriving in the duodenum is promptly neutralized, and only a small segment of duodenum is exposed to acid.

The hormone CCK-PZ is in itself a weak stimulant of pancreatic aqueous secretion, but it exerts a powerful synergistic effect with secretin. CCK-PZ is liberated when fat and protein digestion products come into contact with the duodenal mucosa, and this occurs at the same time acid from the stomach enters the duodenum. Consequently, the two hormones, secretin and CCK-PZ, are liberated at the same time, and acting together, they stimulate secretion of the aqueous component of pancreatic juice. Control of secretion of the aqueous component is exercised through cyclic adenosine monophosphate (cAMP).

Secretion of enzymes is stimulated by cholinergic vagal impulses and by CCK-PZ acting through cyclic guanosine monophosphate (cGMP). When secretion is stimulated, synthesis of enzymes accelerates. The major enzymes secreted are the following.

1. *Proteolytic enzymes.* These include trypsinogen, chymotrypsinogen, procarboxypeptidase and proaminopeptidase. They are all inactive as they are secreted. Enterokinase is an enzyme contained in the brush border of duodenal epithelial cells; it acts on trypsinogen to convert it to the active proteolytic enzyme trypsin. Trypsin in turn activates trypsinogen, chymotrypsinogen and the other proenzymes. Pancreatic juice contains a low concentration of trypsin inhibitor. Trypsin in duodenal contents inhibits release of CCK-PZ, and in that way trypsin is part of a feedback loop regulating pancreatic secretion. The loop is broken when an animal is fed the trypsin inhibitor contained in soybean meal. Because trypsin is inactivated, it no longer inhibits

release of CCK-PZ, and the resulting high concentration of the hormone stimulates abnormal pancreatic secretion and growth.

2. *Amylolytic enzyme.* This is an amylase which catalyzes the hydrolysis of raw or cooked starch and glycogen. The enzyme breaks the α-1,4 glucosidic bonds but not the α-1,6 bond. The result is that the products of starch hydrolysis are glucose, short straight-chain oligosaccharides (maltose and the like) and isomaltose containing a branching point. Pancreatic amylase is continuously released into the plasma across the basolateral borders of the acinar cells, and it is excreted into the urine. Stimulation of the pancreas by cholinergic impulses or by a heavy meal raises plasma amylase concentration, and more severe damage to the pancreas raises it higher.

3. *Lipolytic enzymes.* Pancreatic lipase catalyses the hydrolysis of 1 and 1′ bonds of triglycerides, especially those containing long-chain fatty acids. Bile acids in duodenal contents coat the surface of emulsified fat droplets and thus deny access of pancreatic lipase to its substrate. The pancreas also secretes a protein called *colipase* which binds to the surface of fat droplets and acts as an anchor for lipase, allowing the enzyme to act on triglycerides in the droplets. Pancreatic juice contains other lipases which hydrolyze cholesterol esters and water-soluble esters of fatty acids. The pancreas secretes two phospholipases: A_1 which catalyzes hydrolysis of the 1-bond of phospholipids, and A_2 which catalyzes hydrolysis of the 2-bond. The products are surface-active lysolecithins. Because lecithin is plentiful in bile, there is always a high concentration of lysolecithins in the intestine during digestion.

There are many other enzymes including elastase and ribonuclease in pancreatic juice.

All enzymes are synthesized by ribosomes on the rough endoplasmic reticulum, and they are formed into granules in the Golgi complex. Granules are stored at the apex of a cell, and they are released into the lumen of an acinus when the cell is stimulated by acetylcholine or by CCK-PZ. Each gran-

74 A DIGEST OF DIGESTION

ule apparently contains all the enzymes, but the ratio of one enzyme to another changes throughout the response to a meal. Consequently, the enzymes are not secreted exactly in parallel. Feeding a high fat diet to a rat for several days doubles the concentration of lipase in its pancreatic juice, but there is no evidence that in man the enzyme composition of pancreatic juice changes radically in response to changes in diet.

Pancreatic enzymes are also secreted without being packaged in zymogen granules, and therefore stimulation of secretion of one enzyme need not involve secretion of others. Thus, the candidate hormone chymodenin stimulates secretion of chymotrypsinogen alone.

SUGGESTED READINGS

Code, C. F. (ed.): *Handbook of Physiology: The Alimentary Canal* (Washington, D.C.: American Physiological Society, 1968), Chaps. 52, 57, 122.
Lightwood, R., and Reber, H. A.: Micropuncture study of pancreatic secretion in the cat, Gastroenterology 72:61, 1977.
Malagelada, J-R., DiMagno, E. P., Summerskill, W. H. J., and Go, V. L. W.: Regulation of pancreatic and gallbladder functions by intraluminal fatty acids and bile acids in man, J. Clin. Invest. 58:493, 1976.
Meyer, J. H., and Kelly, G. A.: Canine pancreatic responses to intestinally perfused proteins and protein digests, Am. J. Physiol. 231: 682, 1976.
Palade, G. E.: Intracellular aspects of the process of protein synthesis, Science 189:347, 1975.
Webster, P. D., III, Black, O. J., Mainz, D. L., and Singh, M.: Pancreatic acinar cell metabolism and function, Gastroenterology 73: 1434, 1977.

9. FUNCTIONS OF THE SMALL INTESTINE

The normal human small intestine is several meters long. It is impossible to state its exact length, for it shortens when active and lengthens when quiet. Its length after death is far greater than in life. The length of the small intestine provides a reserve for digestion and absorption. If these functions are impaired, they are spread along more of the small intestine. Only when the intestine is radically shortened, as after surgical removal of a necrotic part, do digestion and absorption become inadequate.

Intestinal epithelial cells are formed by cell division within the crypts, and they migrate up the villi, maturing as they go (Fig 9 – 1). The journey takes 3 days, and at its end, cells desquamate into the lumen. Consequently, between 10 and 25 gm of endogenous protein, together with all other cell constituents, are shed each day, and these are digested and absorbed along with food.

Epithelial cells are held together by tight junctions girdling their apices. Tight junctions differ physiologically from one tissue to another. Those between the surface cells of the stomach normally restrict passage of water and electrolytes; they are very tight junctions. Those in the small intestine are leaky, and they provide a *paracellular pathway* through which flow of water and electrolytes occurs. When cells desquamate, tight junctions are broken, and water, electrolytes and plasma proteins pass into the lumen through the gaps.

The duodenum responds to the chyme delivered to it from the stomach by the following processes.

1. Acid in the chyme is neutralized, and chyme is neutral as it leaves the duodenum. (Chyme becomes more alkaline in the ileum as bicarbonate is secreted by the ileal mucosa.)
   (a) The rate of emptying of the stomach is regulated so that

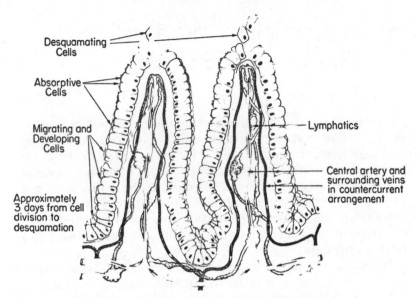

Desquamating Cells

Absorptive Cells

Migrating and Developing Cells

Approximately 3 days from cell division to desquamation

Lymphatics

Central artery and surrounding veins in countercurrent arrangement

Fig 9–1.—The structure of the small intestinal mucosa.

no more acid is emptied into the duodenum than can be neutralized in the duodenal bulb and the first few centimeters of the duodenum.

(b) Secretin and CCK-PZ are released into the blood from the duodenal mucosa, and together they stimulate the pancreas and the liver to secrete bicarbonate-containing fluids.

(c) Some acid is absorbed through the duodenal mucosa or neutralized by bicarbonate in duodenal secretions. In the absence of pancreatic juice, the pH of chyme in the duodenum is about one pH unit lower than normal, which means the chyme is 10 times more acid.

2. The osmotic pressure of duodenal contents is adjusted to isotonicity, and the chyme remains isotonic throughout the rest of the small intestine. Chyme emptied from the stomach may be hypotonic or hypertonic, depending on what is eaten. During digestion of food in the duodenum, a hypertonic solution may be produced. For example, starch has a negligible osmotic pressure, but when hydrolysis is cata-

lyzed by pancreatic amylase many small molecules are rapidly liberated, and these raise the osmotic pressure of duodenal contents.

(a) Gastric emptying is slowed until duodenal contents become isotonic.

(b) Isotonicity is achieved by rapid flow of water from blood into duodenal contents.

3. The major digestion of food begins in the duodenum.

(a) CCK-PZ is released, and it stimulates the pancreas to secrete enzymes. It also stimulates the gallbladder to contract and to empty concentrated bile into the duodenum. The result is that the enzyme and bile acid content of duodenal chyme rises abruptly about 20 minutes after the beginning of a meal.

(b) The contents of the duodenum are thoroughly mixed by segmental movements.

4. Absorption of digestion products begins in the duodenum. Glucose, galactose and other monosaccharides, amino acids and small polypeptides, 2-monoglycerides and free fatty acids are absorbed. As osmotically active particles are absorbed from duodenal contents, a corresponding amount of water is absorbed, and duodenal contents remain isotonic.

Gastric emptying is delayed by hormones and reflexes from the duodenum at the same time acid secretion is inhibited. Acidity and high osmotic pressure of duodenal contents stimulate duodenal motility (Fig 9–2). Vigorous segmental movements spread chyme over a large area, thereby facilitating rapid neutralization and adjustment of osmotic pressure. On the other hand, fat and fat digestion products have little effect upon intestinal motility although they inhibit gastric motility.

Surgical operations on the stomach and duodenum may destroy the duodenum's ability to regulate gastric emptying. For example, a gastrojejunostomy allows contents of the stomach to enter the upper small intestine rapidly. When gastric contents are mixed with pancreatic enzymes, starch and proteins are quickly hydrolyzed, and the contents of the small intestine become hypertonic. Water pours from the blood into the intestinal lumen, the intestine is distended and plasma volume

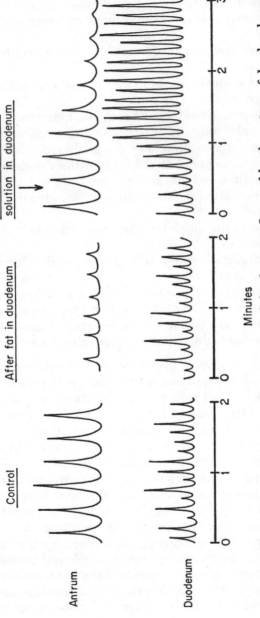

Fig 9–2. — Motility of the human gastric antrum and duodenum influenced by the nature of duodenal contents. The control record shows motility during digestion of a mixed meal. Increases in pressure in the gastric antrum are caused by peristaltic waves whose frequency is three a minute. Irregular increases in pressure in the duodenum occur at a higher frequency. Soon after fat is placed in the duodenum, gastric motility decreases, but small intestinal motility is usually unaffected. Immediately after acid or a hypertonic solution is placed in the duodenum, antral motility decreases, but the strength of duodenal contractions greatly increases. (Adapted from records made by C. F. Code.)

falls. Because the normal means by which gastric emptying is delayed no longer operate, intestinal distention and fall in plasma volume continue unabated. These events, inelegantly termed *dumping*, may be followed by sweating, dizziness and fainting, all the results of a rapid fall in plasma volume.

As chyme enters the duodenum, the motility of the small intestine is stimulated. In contrast with the stomach, the chief movement of the small intestine is segmentation, not peristalsis. In segmentation, the circular muscle of the small intestine contracts in a series of rings, each several centimeters from the other (Fig 9-3). The rings of contraction do not move along the intestine. Chyme is forced from the point of contraction in both directions into the intervening uncontracted segments. Shortly thereafter, the contracted ring of muscle relaxes, and the intervening, hitherto uncontracted segment contracts. Chyme is once more forced in both directions. The process of alternating contraction and relaxation continues indefinitely, with chyme being thrown backward and forward like a flying shuttle.

Segmentation is highly efficient in mixing chyme with digestive juices and in exposing all the chyme to the absorptive surface of the intestinal mucosa. Segmentation also moves chyme slowly down the small intestine, because there is a gradient of frequency of segmentation. The frequency is higher in the upper part of the small intestine than in the lower part. Consequently, chyme is moved from the upper to the lower part, because the chance that any particular part of the chyme will be pushed downward is greater than the chance that it will be pushed upward.

Frequency of segmentation is governed by the frequency of the BER in the local longitudinal muscle. The intrinsic frequency of the BER falls along the length of the small intestine. Response of the circular muscle, as expressed in segmental contraction, is enhanced by presence of chyme in the lumen. Distention of the intestine tends to cause it to contract, and distention of the inactive segment by chyme squirted into it facilitates its subsequent contraction. Segmentation occurs in vagotomized or sympathectomized persons, but it is influ-

Dog's jejunum segmenting

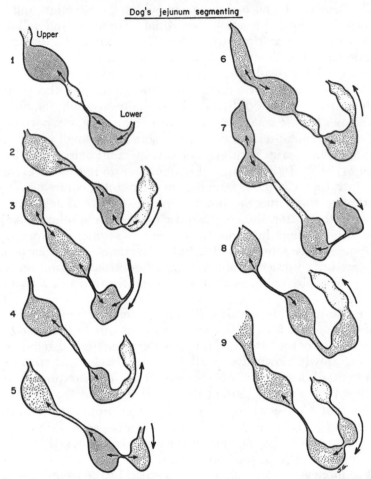

Fig 9–3.—The process of intestinal segmentation. A dog had been fed a meal mixed with the x-ray contrast medium, barium sulfate, and continuous cinefluorographic pictures were taken as the contrast medium entered the upper jejunum. The tracings in this figure were made from the film to show the locus of the medium at approximately 1-second intervals. The *arrows* show the direction in which the jejunal contents were moving at the instant represented by the tracing. Progress of the contents into two distal segments over a period of about 9 seconds shows that intestinal contents can be moved from above downward by segmentation without peristalsis. (Adapted from a film made by H. C. Carlson.)

enced by extrinsic nerves. Vagal stimulation enhances segmentation, and cholinergic drugs are used to promote intestinal movement. Adrenergic and sympathetic influences inhibit
segmentation.

Peristalsis also occurs in the human small intestine, but peristaltic rushes moving a long distance are definitely abnormal
in man. Peristaltic waves occurring during the digestion of a
meal may be no more than a train of two or more peristaltic
contractions slowly moving approximately 10 centimeters
before dying out.

In the interdigestive period, the jejunum is empty (as its
name implies), and the ileum contains a slurry of undigested
fibers and unabsorbed solutes. The terminal ileum is separated from the colon by the ileocecal sphincter. Pressure on the
mucosa of the cecum by cecal contents causes the sphincter to
contract, and distention of the terminal ileum causes it to relax. When a meal is being emptied into the duodenum, the
duodenum begins to segment, and at the same time the ileum
begins to segment as well. This concurrence of gastric and ileal activity is called the *gastroileal reflex*. As the chyme is
pushed into the terminal ileum by segmentation, distention of
the ileum causes the ileocecal sphincter to relax, and with each
terminal ileal segmentation, a small squirt of chyme passes
through the sphincter into the cecum. The cecum itself is
aroused to activity, and the subsequent movements of the colon are said to be caused by the *gastrocolic reflex*. While food
is in the stomach, gastrin is being released. Circulating gastrin
decreases the strength of contraction of the ileocecal sphincter, allowing chyme to pass through it more easily.

When fasting is prolonged, the *interdigestive myoelectric
complex* occurs (Fig 9-4). Strong peristaltic waves sweep
over the gastric antrum, and the duodenum and upper jejunum begin to segment. A wave of vigorous segmentation moves
slowly down the small intestine, and gastric and intestinal
motility die out behind it. Passage from antrum to ileocecal
sphincter takes about an hour, and as one complex reaches the
end of the small intestine, another begins in the gastric antrum
and duodenum. In the dog, and probably in man, feeding
abruptly abolishes the complex.

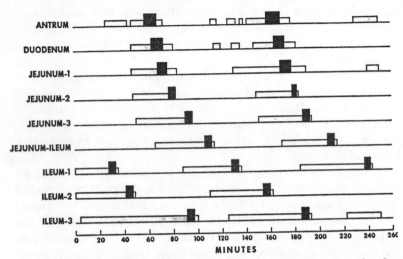

Fig 9–4.—Migrating complexes recorded from the antrum, duodenum, jejunum and ileum of a fasting dog. The *short white areas* represent the periods in which action potentials were beginning to appear and to increase in number in phase with the BER. The *black areas* represent periods of intense activity. The *stippled areas* represent periods in which the activity was dying out. (Adapted from Code, C. F., and Marlett, J. A.: J. Physiol. 246:289, 1975.

The interdigestive myoelectric complex cleans out the small intestine and keeps the bacterial population down. Acid kills bacteria in the stomach and thus protects against enteric infections. The bacterial population of the small intestine is small, usually 10^1 to 10^4 gram-positive aerobes and facultative anaerobes per gram of contents in the jejunum. When intestinal motility is reduced, as in obstruction or in a diverticulum or blind loop, bacteria proliferate rapidly.

In addition to being inhibited by distant events that cause widespread sympathetic adrenergic activity, the intestine is also inhibited by local injury. Pathological dilatation of a part of the small intestine inhibits motility of the part of the intestine above it. This influence of one part of the intestine upon another is called the *intestinointestinal reflex*.

SUGGESTED READINGS

Code, C. F. (ed.): *Handbook of Physiology: The Alimentary Canal* (Washington, D.C.: American Physiological Society, 1968), Chaps. 62, 63, 82, 97, 98, 104, 106, 108, 139.

Code, C. F., and Marlett, J. A.: The interdigestive myo-electric complex of the stomach and small bowel of dogs, J. Physiol. 246:289, 1975.

Daniel, E. E., Gilbert, J. A. L., Schofield, B., Schnitka, T. K., and Scott, G. (eds.): *Proceedings of the Fourth International Symposium on Gastrointestinal Motility* (Vancouver: Mitchell Press, 1974).

Eastwood, G. L.: Gastrointestinal epithelial renewal, Gastroenterology 72:962, 1977.

Vantrappen, G., Janssens, J., Hellmans, J., and Ghoos, Y.: The interdigestive motor complex of normal subjects and patients with bacterial overgrowth of the small intestine, J. Clin. Invest. 59:1158, 1977.

10. ABSORPTION OF WATER AND ELECTROLYTES

In addition to water contained in food and drink, a large volume of water is added to gastrointestinal contents in salivary, gastric, pancreatic and biliary secretions. When a steak meal having an initial volume of 645 ml is eaten, a total volume of 1,500 ml reaches the midjejunum. At the end of the jejunum, 750 ml are left, and in the ileum the volume is reduced to 250 ml. The total daily volume handled by the gut is probably 5 to 10 L, but less than 100 ml are excreted in the stool.

The stomach is relatively impermeable to water, and water does not move readily across the gastric mucosa along osmotic pressure gradients. In contrast, the upper small intestinal mucosa is highly permeable, and osmotic equilibrium is quickly established between contents of the duodenum and blood. Pure water is rapidly absorbed. On the other hand, if the stomach empties a hypertonic solution or if the osmotic pressure of duodenal contents rises as the result of rapid hydrolysis of large molecules within it, water moves into the lumen. Chyme becomes isotonic in the duodenum, and it remains isotonic throughout the rest of the small intestine.

The small intestinal mucosa is permeable to Na^+, and in addition Na^+ is actively reabsorbed. Consequently, net absorption of Na^+ is the result of two opposed unidirectional fluxes (Fig 10–1). The flux from lumen to blood is chiefly the result of active transport of Na^+ by intestinal epithelial cells, and the rate of Na^+ transport is directly proportional to the concentration of Na^+ in luminal fluid. Epithelial cells pump Na^+ from their interior through their lateral borders into interstitial fluid (Fig 10–2). This keeps the concentration of Na^+ low within the cells, and Na^+ diffuses along its concentration gradient from luminal fluid into the cells. When glucose, galactose and

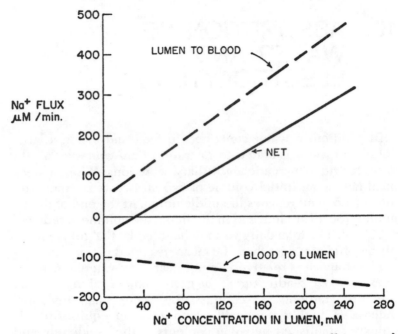

Fig 10–1.—The net absorption of sodium from the small intestinal lumen as a function of the sodium concentration in the lumen. The unidirectional flux of sodium from blood to lumen is very slightly if at all affected by the luminal concentration, whereas the flux from lumen to blood, and consequently the net absorption, increases with increasing luminal concentration. (Adapted from Vaughan, B. E.: Am. J. Physiol. 198:1235, 1960.)

amino acids are present in luminal fluid, they are carried into the cells along with Na^+. Na^+ pumped into interstitial fluid raises the osmotic pressure of the fluid confined between the cells, and water flows through the cells and through the paracellular pathway provided by leaky tight junctions in response to this osmotic pressure gradient. Interstitial spaces expand during Na^+ absorption.

Active Na^+ absorption is thought to be accomplished, in part at least, by exchange of H^+ secreted for Na^+ absorbed. H^+ secreted into the lumen reacts with HCO_3^-. The CO_2

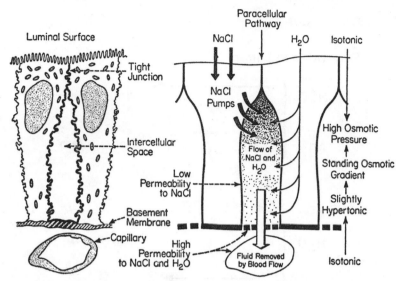

Fig 10–2.—The mechanism of sodium and fluid transport by a layer of epithelial cells such as the mucosa of the small intestine or of the gallbladder. In the intestinal epithelial cells, the sodium and chloride pumps are relatively independent. In each tissue, a standing osmotic pressure gradient established between the lateral borders of the cells is responsible for water flow. (Adapted from Dietschy, J. M.: Gastroenterology 50:692, 1966.)

resulting from this reaction diffuses into blood, and HCO_3^- is in effect absorbed along with Na^+.

Passive flux of Na^+ from blood to lumen apparently occurs chiefly through the paracellular pathway.

There is a gradient of permeability along the length of the intestine: permeability decreases from duodenum to ileum.

Interrelation of water and Na^+ fluxes is shown in Fig 10–3. When solutions containing NaCl are present in the lumen of the small intestine, unidirectional flow of water and Na^+ from blood to lumen is constant and independent of the concentration of luminal fluid. However, unidirectional flux of water from lumen to blood is strongly influenced by the osmotic pressure of the luminal solution. When the solution is hypo-

INTESTiNAL LUMEN   BLOOD

Pure Water In Lumen

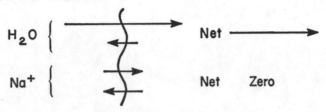

NaCl Up To 210 mN In Lumen

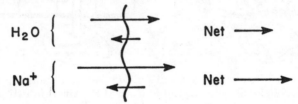

NaCl Above 210 mN In Lumen

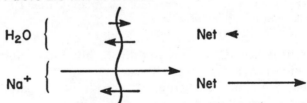

**After NaCl Concentration Is Reduced Below 210mN,
Net Absorption Of Both**

Na^+ And H_2O Occur

Fig 10–3.—The absorption of water and sodium by the small intestine as the result of opposed unidirectional fluxes.

tonic, flux of water from lumen to blood is greater than flux in the opposite direction, and net absorption occurs. As the osmotic pressure of the solution rises, flux of water from lumen to blood decreases until, at a solute concentration approximately equal to a 210 mN solution of NaCl, fluxes in each direction are equal. Above this concentration, flux of water from lumen to blood is smaller than flux in the opposite direction, and the volume of fluid in the lumen increases.

If the concentration of Na^+ in the lumen is greater than about 210 mN, two processes occur at once: net flow of water from blood to lumen and net flow of Na^+ from lumen to blood. Both reduce the concentration of Na^+ in the lumen until its concentration falls below 210 mN. Then net movement of both water and Na^+ is in the direction of lumen to blood, and net absorption of both continues until absorption is complete.

In the lower jejunum and ileum, active absorption of electrolytes and the products of carbohydrate and protein digestion tends to dry out the lumen of the gut, and on account of the relative impermeability of the mucosa, water and electrolytes do not diffuse back into the lumen. However, bile acids and oleic acid, present in the lumen during fat digestion, inhibit Na^+ and water absorption and maintain fluidity of contents of the lower small intestine.

Cl^- can be actively absorbed. If an unabsorbable anion, such as SO_4^{2-}, is present in the lumen, Na^+ required to balance the negative charges cannot be absorbed. Then the concentration of Cl^- is reduced by active absorption almost to zero. In the jejunum and ileum, Cl^- absorption is coupled with HCO_3^- secretion. The concentration of HCO_3^- in the lumen rises to 40–45 mM as the Cl^- concentration falls, and chyme becomes more alkaline as it reaches the terminal ileum.

K^+ is absorbed by passive diffusion throughout the small intestine, and at equilibrium its concentration in intestinal contents is 4–11 mN.

By the time the intestinal contents reach the terminal ileum, digestion and absorption are nearly complete, and fluid containing undigested residues is passed on to the colon. The rate of emptying of the terminal ileum is extremely variable. During the digestive period fluid may enter the colon at the rate of 15 ml per minute. Consequent distention of the cecum stimulates the colon to action, and defecation frequently follows. During the interdigestive period, except when the interdigestive myoelectric complex reaches the terminal ileum about once an hour, no fluid may leave the ileum for the colon. On the average, a total of 1,200–1,800 ml of fluid is delivered to the colon each day. Its average composition is shown in Table 10–1.

The colon normally absorbs Na^+ and secretes K^+, and water is absorbed with Na^+. Absorption of Na^+ is very important in maintaining Na^+ balance, and reabsorption of Na^+ in the colon, but not in the small intestine, is controlled by aldosterone. Although K^+ is secreted by the colon, it is again reabsorbed, apparently by passive diffusion, and the amount of K^+ in the stool is usually far below the daily intake. Nevertheless, diarrhea or frequent colonic irrigation may remove enough K^+ to cause negative K^+ balance.

Diarrhea is defined as fecal water output greater than 500 ml a day. It may occur because the colon's capacity to absorb water and electrolytes is overwhelmed or because the colon's capacity to absorb is reduced. If osmotically active solutes are not absorbed in the small intestine, they carry water with them into the colon. Magnesium and sulfate ions are only slowly absorbed, and consequently ingestion of Epsom salts causes watery diarrhea. Lactose or trehalose may not be absorbed, because lactase or trehalase is absent from the brush border of the intestinal epithelial cells. If so, these sugars are fermented by intestinal flora, and the products cause osmotic diarrhea. The toxins of *Vibrio cholerae* and of some strains of *Escherichia coli* activate adenyl cyclase in intestinal epithelial

TABLE 10–1.—AVERAGE FLUID AND ELECTROLYTE
LOAD AND ABSORPTION IN THE COLON OF
NORMAL HUMAN SUBJECTS°

CONCENTRATION IN TERMINAL ILEAL FLUID		QUANTITY PER 24 HR		
		TERMINAL ILEUM	STOOL	ABSORBED
Water		1,500 ml	40 ml	1,460 ml
Na^+	127 mN	196 mEq	1 mEq	195 mEq
K^+	6 mN	9 mEq	5 mEq	4 mEq
Cl^-	67 mN	103 mEq	1 mEq	102 mEq

°Adapted from Phillips, S. F., and Geller, J.: J. Lab. Clin. Med. 81:733, 1973.

cells, and cAMP in turn stimulates the cells to secrete an enormous volume of fluid similar to an ultrafiltrate of plasma. The volume is too large to be reabsorbed in the colon, and cholera and traveler's diarrhea result in potentially fatal dehydration.

The ability of the colon to absorb water and electrolytes may be reduced. Some bile acids inhibit colonic absorption of Na^+. Under normal circumstances, these bile acids do not reach the colon in significant amount, but when the ileum is diseased or resected, enough may enter the colon to cause diarrhea. Hydroxylated fatty acids such as ricinoleic acid, the active ingredient of castor oil, inhibit salt and water absorption. Unsaturated fatty acids are hydroxylated by intestinal and colonic bacteria, and when there are errors in fat digestion or absorption, hydroxylated fatty acids may be produced in amounts sufficient to cause diarrhea.

SUGGESTED READINGS

Code, C. F. (ed.): *Handbook of Physiology: The Alimentary Canal* (Washington, D.C.: American Physiological Society, 1968), Chaps. 63, 65, 66, 74, 76, 136.

Fordtran, J. S.: Stimulation of active and passive absorption by sugars in the human jejunum, J. Clin. Invest. 55:728, 1975.

Love, A. H. G., Rohde, J. E., Abrams, M. E., and Veal, N.: The measurement of bidirectional fluxes across the intestinal wall in man using whole gut perfusion, Clin. Sci. 44:267, 1973.

Turnberg, L. A., Bieberdorf, F. A., Morawski, S. G., and Fordtran, J. S.: Interrelationships of chloride, bicarbonate, sodium, and hydrogen transport in the human ileum, J. Clin. Invest. 49:557, 1970.

Turnberg, L. A., Fordtran, J. S., Carter, N. W., and Rector, F. C., Jr.: Mechanism of bicarbonate absorption and its relationship to sodium transport in the human jejunum, J. Clin. Invest. 49:548, 1970.

Wilson, T. H.: *Intestinal Absorption* (Philadelphia: W. B. Saunders Co., 1962).

Wiseman, G.: *Absorption from the Intestine* (New York: Academic Press, 1964).

11. ABSORPTION OF CALCIUM AND IRON

About 1,600 mg of calcium contained in the diet and 600 mg contained in secretions and desquamated cells enter the gut lumen each day. Of this, 700 mg are absorbed, leaving 1,500 mg to be excreted in the stool. Although 700 mg are absorbed, net absorption is only 100 mg a day.

When the concentration of free calcium in the lumen is below 5 mM, absorption is accomplished entirely by an active process. When the concentration is above 5 mM, additional calcium is absorbed by passive diffusion.

Calcium is absorbed against a concentration gradient by first binding to a specific carrier protein in the brush border of intestinal epithelial cells and then being transported through the cells to be discharged into interstitial fluid. The calcium-binding protein is absent in rachitic animals who lack vitamin D. When vitamin D_3 is given, the vitamin is converted to 1,25-dihydroxy-vitamin D_3, and within 90 minutes the calcium-binding protein appears in the intestinal mucosa. Parathyroid hormone does not increase calcium binding, but it promotes calcium release from mucosal cells to interstitial fluid.

Net calcium absorption increases when dietary calcium is increased. It is zero when calcium intake is 0.1 mM (4 mg) per kg body weight, and it rises to a maximum of 3 mM (120 mg) per kg a day when intake is high. Calcium absorption is facultatively regulated to meet body needs. The ability to absorb calcium is increased by calcium deprivation. Young and growing persons absorb calcium more rapidly than do mature or elderly ones. Lactating women absorb calcium avidly.

Bile acids affect calcium absorption in two ways. Vitamin D is a fat-soluble vitamin, and in bile acid deficiency micelle formation and, therefore, vitamin D absorption are reduced. Free fatty acids form insoluble calcium soaps. When steatorrhea occurs as result of bile acid deficiency, calcium is carried into the stool by free fatty acids. This results in negative calcium balance and osteomalacia.

When calcium is sequestered in calcium soaps, it is unavailable to precipitate oxalic acid eaten in oxalate-containing foods such as rhubarb. Consequently, in persons with steatorrhea, oxalic acid is absorbed 5 times more rapidly than in normal persons, and because oxalic acid is excreted in the urine together with calcium, such patients often have calcium oxalate kidney stones.

Magnesium is apparently absorbed only by passive diffusion.

Most iron released at the end of their life by erythrocytes is recycled, but iron absorption must keep up with iron loss. Normal iron intake is 15 to 20 mg a day, most of it in hemoglobin and myoglobin. If iron intake is increased, particularly by adding inorganic iron to the diet, iron absorption increases.

Iron absorption increases as the need for iron increases. A man absorbs about 0.5 to 1 mg a day, but a menstruating woman, on account of her chronic mild anemia, absorbs more. The rate of absorption increases following hemorrhage, but not for 3 days. The reason for the delay is that the message of iron deficiency is given to cells in the crypts as they are being formed by cell division, but the cells do not absorb iron at an increased rate until they have migrated to the tips of the villi 3 days later.

Iron can be absorbed along the whole length of the intestine, but efficiency of absorption is greatest in the duodenum and upper jejunum. Heme iron is more readily absorbed than inorganic iron. Heme is absorbed as such into the mucosal cells, and in them iron is split from the porphyrin by heme oxidase. The nature of the intestinal contents influences absorption of inorganic iron. Iron bound to phosphoproteins and precipitated in calcium phosphates is unavailable for absorption. Ferric iron, because it forms insoluble hydroxides at the pH of intestinal contents, is less readily absorbed than ferrous iron. Ascorbic acid reduces ferric to ferrous iron and promotes absorption. By chelating with iron, ascorbic acid also releases iron bound to other compounds and makes it available for absorption. Wine does not increase iron absorption.

Inorganic iron moves in both directions across the apical membrane of intestinal epithelial cells: it is actively absorbed and actively secreted. In the duodenum, absorption predominates over secretion, but secretion may predominate in lower parts of the intestine.

Iron is absorbed by a specific iron-binding protein. Once inside the mucosal cell, iron is partitioned into at least two pools. Still combined with a specific protein, some iron enters an absorbable pool and is carried through the cell to be extruded into the interstitial fluid and plasma. In plasma, it is carried by another iron-binding protein, transferrin (Fig 11–1). Absorption and transport are increased in iron deficiency. The rest of the iron enters a pool which is either slowly absorbed or is entirely unabsorbable. Iron in that pool is lost into the lumen when cells containing it desquamate.

Iron also enters intestinal epithelial cells from the plasma, and that iron can be secreted by the cells into the lumen. En-

Fig 11–1.—Absorption and excretion of iron in the small intestine.

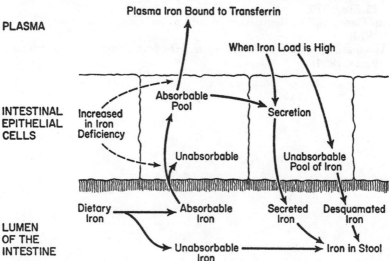

try from the plasma and secretion into the lumen are increased when the iron load is high. Iron entering epithelial cells from plasma is also lost when the cells desquamate. Consequently, iron in the stool consists of unabsorbed and unabsorbable iron, recently absorbed iron contained in desquamated cells, iron secreted by epithelial cells and iron derived from the plasma and either secreted or lost as cells desquamate. Of course, iron contained in the cells' own cytochrome and other iron-containing enzymes is also lost when the cells are shed.

Iron absorption is inhibited by both cobalt and manganese which compete with iron for one step in the process of absorption.

SUGGESTED READINGS

Code, C. F. (ed.): *Handbook of Physiology: The Alimentary Canal* (Washington, D.C.: American Physiological Society, 1968), Chap. 78.

Ireland, P., and Fordtran, J. S.: Effect of dietary calcium and age on jejunal absorption in humans studied by intestinal perfusion, J. Clin. Invest. 52:2672, 1973.

Kretsinger, R. H.: Calcium-binding proteins, Ann. Rev. Biochem. 45:239, 1976.

van Campen, D.: Regulation of iron absorption, Fed. Proc. 33:100, 1974.

Wiseman, G.: *Absorption from the Intestine* (New York: Academic Press, 1964).

12. DIGESTION AND ABSORPTION OF CARBOHYDRATES

Carbohydrates make up 50–60% of the diet, 250–800 gm a day providing 1,000–2,500 kilocalories. The major carbohydrates are starches, glycogen and disaccharides. Starches contain long chains of glucose molecules linked by α-1,4 glucosidic bonds. Branches originate at α-1,6 bonds (Fig 12–1). The chief disaccharides are sucrose, consisting of glucose and fructose, and lactose, consisting of glucose and galactose. There are many other carbohydrates in the diet including pentoses and disaccharides such as trehalose of mushrooms.

Hydrolysis of starch occurs in the lumen of the stomach and small intestine, and hydrolysis is catalyzed by salivary and pancreatic amylase. These enzymes attack only the α-1,4 glucosidic bonds, and the products of hydrolysis of this bond are glucose, maltose (glucose-glucose), some trisaccharides and tetrasaccharides, all containing the α-1,4 bond, and branched oligosaccharides containing the α-1,6 branching point. Starch is quickly hydrolyzed in the duodenum, and a few large molecules are broken into many small ones. As a result, osmotic pressure of duodenal contents tends to rise during digestion of a large starchy meal.

Further hydrolysis of disaccharides and other oligosaccharides is catalyzed by enzymes confined to the brush border of intestinal epithelial cells. Among these are several maltases, which catalyze hydrolysis of the α-1,4 bonds, and isomaltase, which catalyzes hydrolysis of the α-1,6 bond at the branching points. Other enzymes of the brush border include sucrase (or invertase), acting on sucrose to give glucose and fructose, and lactase, acting on lactose to give glucose and galactose.

The products of hydrolysis are absorbed in three ways.

97

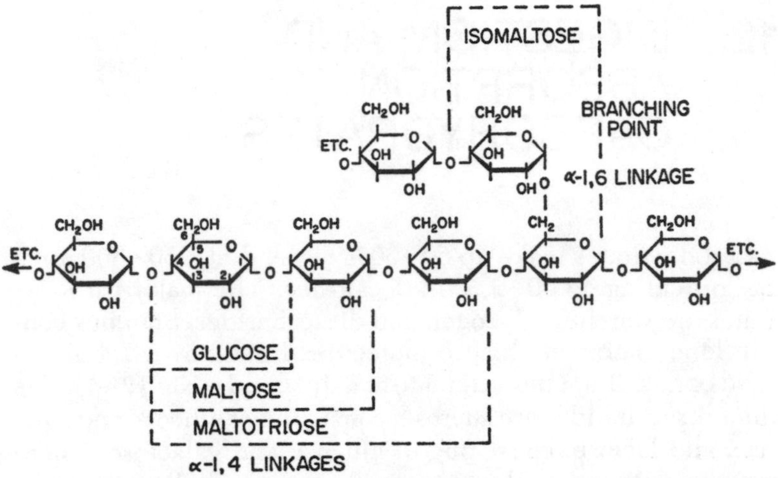

Fig 12-1.—The structure of starch. Hydrolysis catalyzed by pancreatic amylase occurs at the α-1,4 linkage, and the products of hydrolysis are straight-chain oligosaccharides. Since pancreatic amylase does not catalyze hydrolysis of the α-1,6 branching point linkage, isomaltose is also a product of hydrolysis. Further hydrolysis is catalyzed by the maltases and the isomaltase of the brush border of the intestinal epithelial cells.

1. Some monosaccharides are absorbed by diffusion, and the rate of absorption is directly proportional to the concentration of the sugar in the lumen. Fructose is an example (Fig 12-2).
2. Glucose and galactose are rapidly absorbed by active transport. They share the same transport mechanism, and consequently they compete for absorption. The transport process requires presence of Na^+ in the lumen, and Na^+ is absorbed along with glucose and galactose (Fig 12-3). In order to cross the brush border, each Na^+ attaches to a carrier contained in the border. A glucose or galactose molecule attaches to the same carrier and is transported into the cell along with Na^+. The concentration of Na^+ within the cell is kept low by a sodium pump which extrudes Na^+ into the interstitial fluid. The diffusion gradient for Na^+ from lumen

to cell provides the energy necessary to transport the hexose into the cell. The pentose, xylose, is also absorbed by a Na^+-dependent process.

Coupling of glucose and Na^+ absorption is useful in the management of cholera. In that disease, a large volume of fluid containing Na^+ at 140 mN is secreted by the mucosa of the jejunum and ileum. Diarrhea at the rate of 600 ml per hour may occur, and death quickly follows as the result of collapse of extracellular fluid volume. If the patient is given an isotonic Na^+-containing fluid to drink, the fluid is not absorbed. If, however, glucose is added to the fluid, all is absorbed, and extracellular fluid volume is sustained. Su-

Fig 12–2.—Mean rates of absorption of glucose and fructose in the jejunum of normal subjects and in patients with gluten enteropathy. The sugars were delivered in the concentrations shown by perfusion of a 30-cm length of jejunum at the rate of 20 ml per minute. Glucose absorption in the normal subjects appears to be by way of a saturable transport system. Fructose absorption in normal subjects appears to be by way of diffusion only. (Adapted from Holdsworth, C. D., and Dawson, A. M.: Gut 6:387, 1966.)

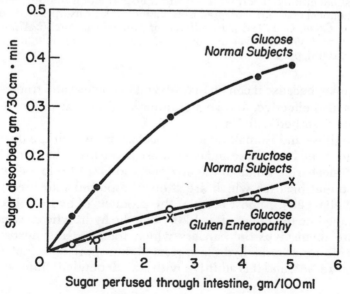

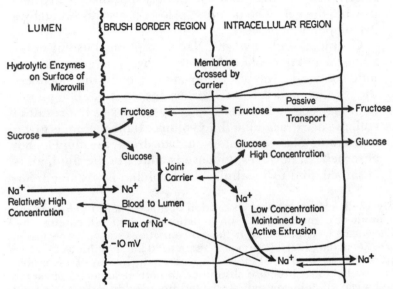

Fig 12–3.—Scheme showing coupling of glucose absorption by the intestinal epithelial cell with sodium transport along sodium's electrochemical gradient. (Adapted from Crane, R. K.: Absorption of Sugars, in Code, C. F. [ed.]: *Handbook of Physiology:* Sec. 6. *Alimentary Canal,* Vol. III [Washington, D. C.: American Physiological Society, 1968], pp. 1323–1351.)

crose, because it can be hydrolyzed to glucose and fructose, is also effective, and so are amino acids, because they too are absorbed with Na^+.

3. Maltose and isomaltose are disaccharides which can be hydrolyzed by maltase and isomaltase in the brush border. The disaccharide binds to the enzyme and is split into its component hexoses which are then transported into the epithelial cells. Almost none of the products of hydrolysis escape into the lumen. It is likely that the hydrolytic enzymes are themselves the carriers which transport the hexoses. This process of hydrolase-related transport does not require Na^+, and it is additive with Na^+-dependent transport.

Actively transported sugars accumulate within the epithelial cells, and they apparently diffuse passively from the cells to interstitial fluid. Within the cells some glucose is used for energy, and some is converted to glycerol phosphate to be used in synthesis of triglycerides and phospholipids. Intestinal epithelial cells have a high rate of aerobic glycolysis, and the lactate they produce is transported into interstitial fluid.

The oral glucose tolerance test in which a person is given a solution of 50 gm of glucose to drink is as much a test of his ability to absorb glucose as it is of his ability to handle glucose once it is absorbed. In the rare patient whose glucose-absorbing system is defective, the concentration of glucose in his blood does not rise after ingestion of glucose; his blood sugar curve is flat. The curve may also be flat in patients who are allergic to milk proteins.

Constituent hexoses of disaccharides cannot be absorbed until the disaccharide is hydrolyzed by a specific enzyme in the brush border. The enzyme most commonly absent is lactase. Although lactase is usually present in the brush border at birth, it frequently disappears at the age of 2–6 years. Adult lactase deficiency is genetically determined. About 70% of adult American blacks are lactase-deficient, and other groups with high incidence of lactase deficiency include Ashkenazic Jews, Arabs, Greek Cypriots, Japanese, Formosans and Filipinos. Only 5–15% of adult white Americans of Northern European ancestry lack the enzyme. In the absence of lactase, lactose is unabsorbed, and lactose taken in milk is fermented in the small intestine and colon. Copious osmotic diarrhea follows, and this may be fatal in lactase-deficient babies. Switching from lactose to some other sugar circumvents the problem, and adults avoid the consequences of lactose intolerance by not drinking milk. If a lactase-deficient person drinks a solution of lactose, his blood sugar curve is flat.

Other oligosaccharidase deficiencies have been recognized. A combined deficiency of isomaltase and sucrase has been identified, and a deficiency of trehalase, which results in diarrhea when its victims eat mushrooms, is a rare medical oddity.

SUGGESTED READINGS

Code, C. F. (ed.): *Handbook of Physiology: The Alimentary Canal* (Washington, D.C.: American Physiological Society, 1968), Chaps. 69, 118, 119.

Bayless, T. M., Rothfield, B., Massa, C., Wise, L., Paige, D., and Bedine, M. S.: Lactose and milk intolerance: Clinical implications, N. Engl. J. Med. 292:1156, 1975.

Gray, G. M.: Carbohydrate digestion and absorption, Gastroenterology 58:96, 1970.

Ramaswamy, K., Malathi, P., Caspary, W. F., and Crane, R. K.: Characteristics of the disaccharide-related transport system, Biochim. Biophys. Acta 345:39, 1974.

Wilson, T. H.: *Intestinal Absorption* (Philadelphia: W. B. Saunders Co., 1962).

Wiseman, G.: *Absorption from the Intestine* (New York: Academic Press, 1964).

13. DIGESTION AND ABSORPTION OF PROTEIN

Under normal circumstances, most dietary protein is completely digested and absorbed. The protein of the stool is that contained in bacteria, desquamated cells and the mucoproteins of colonic secretions. In states of maldigestion and malabsorption that cause steatorrhea, the loss of dietary protein closely parallels the loss of fat.

In the adult, a small amount of protein is absorbed intact through intestinal cells by the process of pinocytosis. An example of absorption of a large protein-containing molecule is furnished by the absorption of vitamin B_{12} and intrinsic factor in the terminal ileum. Intrinsic factor is a glycoprotein having a molecular weight between 45,000 and 50,000. It is secreted by the oxyntic cells of the stomach. It combines with vitamin B_{12} of the diet, and in doing so it forms dimers and trimers. The complex of vitamin B_{12} and intrinsic factor is absorbed intact into the epithelial cells of the ileum, and the vitamin is then transferred to other carriers in the plasma. The amount of other native proteins absorbed is nutritionally negligible, but allergic reactions to absorbed native proteins can be the basis of food sensitivities.

Mammalian fetuses do not synthesize their own antibodies, and in some species, including man, passive immunity at birth results from placental transfer of maternal antibodies. In many other species, including ruminants and rodents, placental transfer does not occur, and the plasma of their newborn is devoid of γ-globulin. In those species, the whey of maternal colostrum contains IgG antibodies, and these are ingested by the suckling young. During the first 36 hours of life, IgG antibodies are absorbed intact through the intestinal mucosa by pinocytosis.

103

A large amount of endogenous protein is digested each day and absorbed as amino acids and polypeptides. There are three sources.

1. About 10 gm of protein enzymes are secreted into the intestinal tract each day. Most is digested and absorbed, but some pancreatic enzymes can be found in the stool.
2. There is a very rapid turnover of cells of the gastrointestinal epithelium, and the whole mucosa is replaced in about 3 days. This means that between 100 and 250 gm of mucosal cells are shed into the lumen each day, and these contribute 10 or more gm of protein to be digested and absorbed.
3. Under normal circumstances, a small amount of plasma proteins leaks into the digestive tract each day. The amount of albumin lost this way is about 1–4 gm a day, and a corresponding amount of other plasma proteins is probably lost into the gut as well. In states of protein-losing gastroenteropathy, an enormous amount of plasma proteins may be shed into the intestinal tract. In one instance, the proteins contained in 5 L of plasma were found to be lost through the gastric mucosa in one day. Plasma proteins are digested along with other proteins, and their constituent amino acids are returned to the liver to be used for resynthesis of plasma proteins. In severe protein-losing states, however, the maximum synthetic ability of the liver is incapable of maintaining the normal plasma protein concentration, and plasma albumin concentration falls severely.

Eight pepsinogens, differing slightly in their properties, are secreted by the chief cells of the stomach. They are converted into the active proteolytic enzymes, the corresponding pepsins, by acid and by pepsin itself. Because gastric contents become acid only well after digestion has begun, only a fraction of the protein in the diet is attacked by pepsins in the stomach. Gastric digestion is unnecessary; persons secreting no acid or pepsinogens digest protein adequately.

Proteolytic enzymes are secreted in pancreatic juice as inactive precursors, and they are activated in the lumen by enterokinase and by trypsin. Trypsin hydrolyzes only peptide bonds in which the carbonyl fraction is lysine or arginine, and therefore products of tryptic digestion are polypeptides of various lengths. Chymotrypsin is slightly less specific. Aminopeptidases and carboxypeptidases liberate amino acids from the ends of polypeptide chains. Pancreatic enzymes do not hydrolyse dipeptides.

Within 15 minutes after protein is emptied from the stomach, as much as 30–50% of it is broken down to small peptides and amino acids. Although luminal hydrolysis is rapid, complete digestion and absorption of 50 gm of protein may take 4–6 hours, and this occurs throughout the whole small intestine.

Complete hydrolysis of proteins is not necessary for their absorption, because small polypeptides as well as individual amino acids are absorbed by cells at the tips of the intestinal villi. Polypeptides containing two or three amino acids are, in fact, absorbed more rapidly than the corresponding free amino acids. Di- and tripeptide absorption does not require the cooperation of Na^+. Some polypeptides are hydrolyzed to amino acids by peptidases which are an integral part of the brush border, and their free amino acids are immediately transported into the cells. Other polypeptides enter the cells intact and are hydrolyzed by peptidases contained in the cytosol. A small fraction of the polypeptides absorbed appear intact in portal blood along with free amino acids. Because intestinal cells are desquamated, their peptidases contribute to hydrolysis of polypeptides within the lumen of the jejunum and ileum.

Free amino acids are absorbed by at least three active transport systems.

1. Neutral amino acids are absorbed by one transport system, and as a result they compete with one another. The system is not absolutely specific for L-amino acids.
2. Basic amino acids are carried by a second system at a slower rate.
3. A third system transports proline and hydroxyproline.

The amino acid transport systems require the presence of Na^+ in the lumen, and the mechanism of transport may be similar to that of glucose and galactose.

Amino acids accumulate in epithelial cells during absorption, and some are metabolized. Glutamic and aspartic acids undergo transamination with pyruvic acid, forming alanine, and glutamine is oxidized. There is only a small rise in the concentration of amino acids in plasma during digestion of a meal, for amino acids are disposed of by tissues almost as quickly as they are released from intestinal epithelial cells.

In normal persons, some nitrogenous compounds escape into the colon, where they are attacked by colonic bacteria. Ammonia liberated by deamination is absorbed by diffusion into portal blood, and upon reaching the liver it is used for urea synthesis. In persons with liver disease or with portacaval shunt, the absorbed ammonia fails to be quickly removed by the liver, and the consequent rise in blood ammonia may cause encephalopathy.

In a patient whose renal tubules are unable to reabsorb some amino acids, there may be a small deficiency in intestinal absorption as well. A person with cystinuria does not absorb arginine or lysine in his intestine as rapidly as a normal person. One whose renal tubules fail to reabsorb threonine and tryptophan has reduced intestinal absorption of tryptophan; he may develop pellagra-like symptoms correctable by feeding nicotinic acid. The consequences of intestinal defects are much smaller than those of the corresponding renal defects, because the defects are confined to the absorption of free amino acids. Amino acids in small peptides are absorbed, and a patient who cannot absorb phenylalanine or tryptophan in the free state can absorb them when they are present in dipeptides.

Unabsorbed amino acids are degraded in the colon to potentially toxic substances such as cadaverine and putrescine, and absorption of these through the colonic mucosa may lead to neurological abnormalities.

Some persons in temperate climates have a collection of absorption defects called *nontropical sprue*—in infants, *celiac disease.* Because the disease can be traced to the effects of gluten from wheat, barley or rye flour, it is also called *gluten enteropathy.* The intestinal mucosa is flat, and villi are absent. The microvilli are clubbed and short, and digestive enzymes normally present in the brush border are absent or greatly reduced in concentration. Absorption of all nutrients is impaired, and fermentation of food in the gut leads to bloating and diarrhea. The condition can be relieved by complete elimination of gluten from the diet, and when a patient in remission is challenged by being fed gluten, the signs and symptoms promptly reappear. Some persons believe the disease is the result of an immunological process, because many, but not all, patients have antibodies to gluten in their blood. Others think that a metabolic defect allows undigested and toxic fragments of gluten to accumulate in the epithelial cells.

SUGGESTED READINGS

Code, C. F. (ed.): *Handbook of Physiology: The Alimentary Canal* (Washington, D.C.: American Physiological Society, 1968), Chaps. 67, 68, 75, 118, 120, 121, 122, 133, 138.

Gray, G. M., and Cooper, H. L.: Protein digestion and absorption, Gastroenterology 61:535, 1971.

Matthews, D. M.: Intestinal absorption of peptides, Physiol. Rev. 55: 537, 1975.

Matthews, D. M.: Peptide absorption — then and now, Gastroenterology 73:1267, 1977.

Matthews, D. M., and Adibi, S. A.: Peptide absorption, Gastroenterology 71:151, 1976.

Silk, D. B. A., Nicholson, J. A., and Kim, Y. S.: Hydrolysis of peptides within the lumen of the small intestine, Am. J. Physiol. 231:1322, 1976.

Walker, W. A., and Isselbacher, K. J.: Uptake and transport of macromolecules by the intestine. Possible role in clinical disorders, Gastroenterology 67:531, 1974.

Wiseman, G.: *Absorption from the Intestine* (New York: Academic Press, 1964).

14. BILE ACIDS

The major constituents of the bile are these:

1. A solution of sodium bicarbonate and chloride the secretion of which is independent of the secretion of bile acids, and which is controlled by the hormones secretin and cholecystokinin-pancreozymin. This component has been described in the section on pancreatic secretion (Fig 14 – 1).
2. Bile acids, primary and secondary, almost entirely conjugated.
3. An electrolyte solution accompanying the bile acids, the rate of secretion of which is governed by the rate of secretion of bile acids.
4. Lecithin.
5. Cholesterol, which, together with lecithin and bile acids, is chiefly contained in micelles but which may be in the form of microcrystals in hepatic bile.
6. Bile pigments, chiefly bilirubin conjugated with glucuronic acid.
7. Some protein.
8. Many compounds metabolized and secreted by the liver: detoxified drugs, phenolsulfonphthalein (PSP) in conjugated form, etc., etc.

Bile acids are steroid derivatives of cholesterol (Fig 14–2). Their chemical features are the steroid nucleus to which 3, 2 or 1 hydroxyl groups are attached and a short side chain ending in a carbonyl group. Primary bile acids are those which are synthesized by the liver; these are cholic acid with three hydroxyl groups and chenodeoxycholic acid with two hydroxyl groups. Primary bile acids may be modified by bacterial action in the small intestine and colon, and the products are called secondary bile acids. The most important secondary bile acids are deoxycholic acid, derived from cholic acid which has two

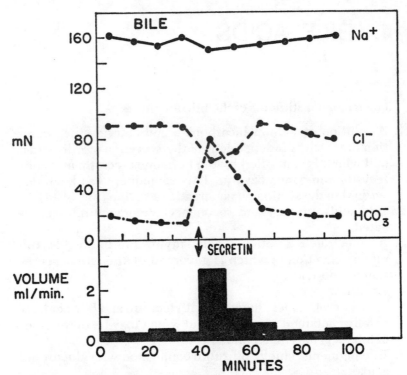

Fig 14–1.— The composition of the bile acid-independent fraction of human bile. Secretion of an isotonic, bicarbonate-rich fluid is stimulated by secretin administered intravenously or released from the duodenal mucosa. (Adapted from Waitman, A. M., et al.: Gastroenterology 56:286, 1969.)

hydroxyl groups, and lithocholic acid, which has only one hydroxyl group. When secondary bile acids are absorbed and returned to the liver in portal blood, they are secreted into bile along with the primary bile acids.

The liver conjugates bile acids with glycine or taurine by forming a peptide bond between the carbonyl group on the side chain of the bile acid and the amino group of glycine or taurine. If the bile acid is cholic acid, the conjugates are called glycocholic and taurocholic acid, respectively. Because the

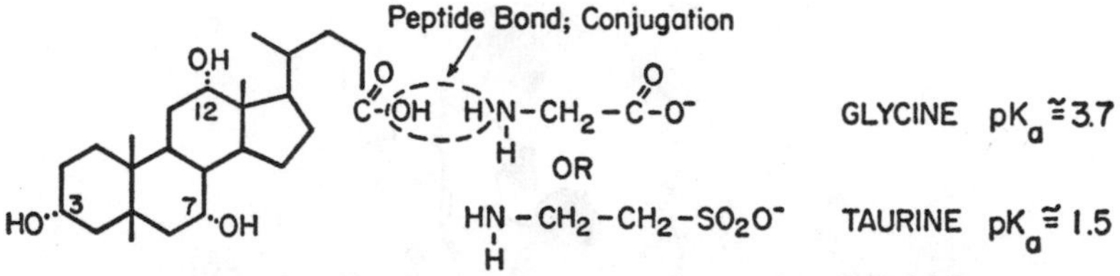

Peptide Bond; Conjugation

CHOLIC ACID (Primary Bile Acid)

GLYCINE $pK_a \cong 3.7$

TAURINE $pK_a \cong 1.5$

NO — OH on 12 = Chenodeoxycholic Acid (primary)
NO — OH on 7 = Deoxycholic Acid (secondary)
NO — OH on 7 and 12 = Lithocholic Acid (secondary)

Fig 14–2.—The structure of the common bile acids. The bile acid is conjugated with either glycine or taurine by elimination of water to form a peptide bond. The approximate ionization constants of glycocholic and taurocholic acids are given on the right. (Adapted from Hofmann, A. F.: Gastroenterology 48:484, 1965.)

bond between cholic acid and taurine is a peptide bond, the compound can be called cholyltaurine. Conjugates of other bile acids have corresponding names. In man, the ratio of glycoconjugates to tauroconjugates is about 3 to 1. The reason is that taurine is in relatively short supply, whereas that of glycine is unlimited. If taurine is fed, more tauroconjugates are secreted. If taurine is lost by failure to reabsorb conjugated bile acids, the ratio of glycoconjugates to tauroconjugates may rise to 20 to 1. Taurine is a stronger acid than glycine, and consequently a greater fraction of tauroconjugates is ionized at the pH of intestinal contents.

The total amount of bile acids, primary and secondary, conjugated and unconjugated, in the body at any one time is called the *bile acid pool,* and in a normal man the pool size is about 2–4 gm.

A conjugated bile acid is a flat molecule, water-soluble on one side and fat-soluble on the other (Fig 14–3). The hydrophilic hydroxyl groups, the peptide bond and the terminal group of glycine or taurine project to one side, and the other side is made up of hydrophobic groups. Consequently, bile acids accumulate at oil-water interfaces, and they stabilize the interface.

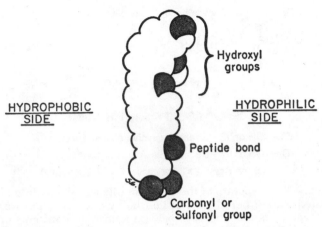

Fig 14–3.—A conjugated cholic acid molecule (side-on view). (Adapted from D. M. Small.)

All bile acids, whether conjugated or not, returning to the liver in portal blood are promptly secreted into the bile (Fig 14–4, A). Unconjugated bile acids are conjugated before they are secreted. A small amount of unconjugated bile acids is present in bile only when massive amounts of unconjugated bile acids reach the liver. Bile acids are always secreted in conjunction with lecithin and cholesterol. These three compounds aggregate in micelles. Both bile acids and lecithin are polar molecules; in a micelle, their fat-soluble portions form a hydrophobic core and their water-soluble portions form a hydrophilic shell. Cholesterol, which is almost totally insoluble in water, dissolves in the hydrophobic core of the micelle.

Bile is a neutral solution, and the bile acids are ionized. Their negative charges are balanced by cations, chiefly sodium, and these form a counterion shell around the micelles. Osmotic pressure is determined by the total number of particles in solution, and bile is isotonic with plasma. Because the organic constituents of bile, together with the cations, aggregate in micelles, the total number of particles in solution is far less than the total number of particles determined by chemical analysis. The osmotic coefficient of Na^+ in bile is only about

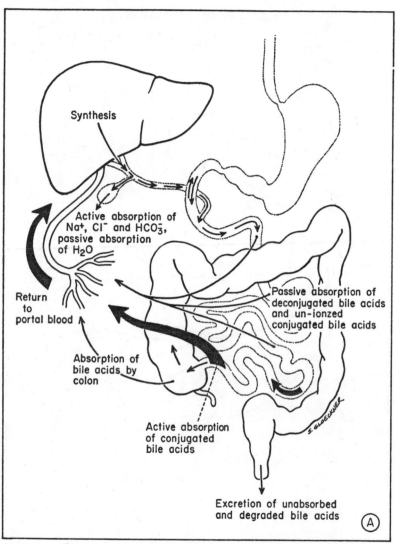

Fig 14–4.—The enterohepatic circulation of bile acids. **A**, the complete system: synthesis in the liver, secretion of newly synthesized and reabsorbed bile acids, storage in the gallbladder, passive and active absorption in the small intestine, escape of a small fraction into the colon, passive absorption in the colon and excretion into the feces, and return of absorbed bile acids to the liver in the portal blood. (Continued.)

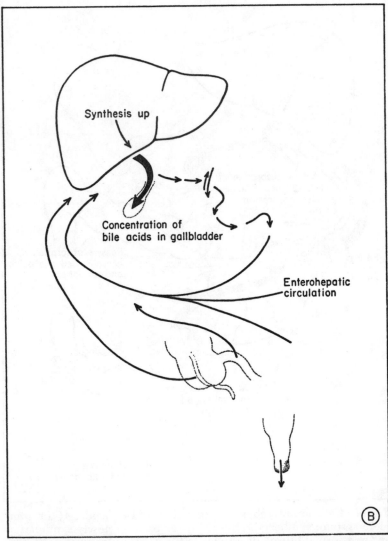

Fig 14–4 (cont.).—B, the interdigestive phase: secretion and storage in the gallbladder, minimal bile acids in intestine with minimal absorption and return to the liver, increased synthesis as a result of reduced negative feedback control. (Continued.)

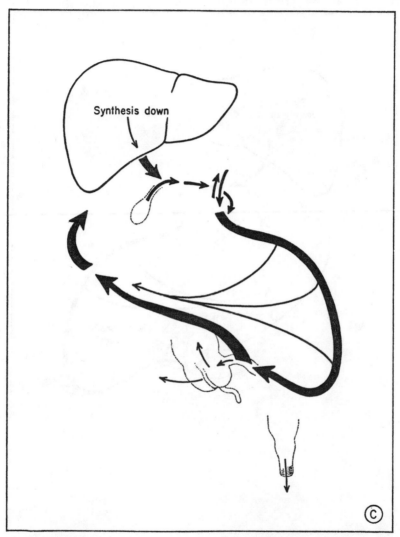

Synthesis down

Fig 14–4 (cont.).— **C,** the digestive phase: emptying of the gallbladder, large amounts of bile acids in the intestine with rapid absorption, rapid secretion of reabsorbed bile acids, reduced synthesis on account of increased negative feedback control. (Continued.)

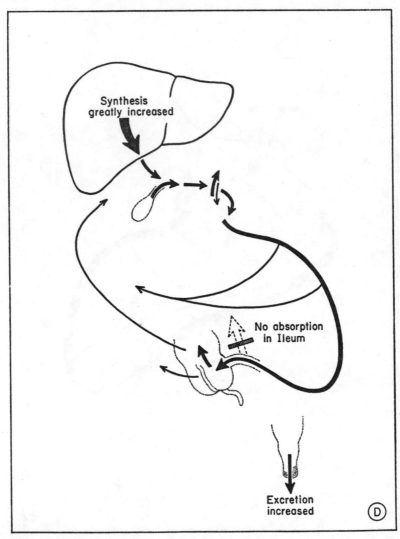

Fig 14–4 (cont.). – D, interruption of active reabsorption in the terminal ileum: decreased or absent active reabsorption, decreased return
to liver with greatly increased synthesis, increased escape of bile
acids into colon and feces.

0.3. Any increase in the rate of bile acid secretion is accompanied by an increase in the volume flow of bile. This is called the *choleretic* effect of bile acids.

During the interdigestive period, the gallbladder is relaxed, and resistance of the sphincter separating the common bile duct from the duodenum is relatively high (Fig 14–4, B). Only a small fraction of the bile secreted by the liver flows into the intestine; the rest is diverted to the gallbladder. The mucosa of the gallbladder actively reabsorbs Na^+, Cl^- and HCO_3^- ions from gallbladder bile, and as these are reabsorbed, water passively follows. The result is that at the end of the interdigestive period the gallbladder contains 40–60 ml of an isotonic solution in which the bile acid-lecithin-cholesterol micelles are highly concentrated. Bile acids may be 5 to 10 times more concentrated in gallbladder bile than in hepatic bile. Because bile pigments are concentrated as well, gallbladder bile is almost black (hence, *melancholia*, a disorder of black bile).

Soon after the beginning of a meal, the first chyme to reach the duodenum stimulates release of cholecystokinin from the duodenal mucosa. The hormone causes the gallbladder to contract, and over the next 20 or so minutes the gallbladder's store of concentrated bile enters the duodenum (Fig 14–4, C). Bile acids travel down the intestine with the chyme and are reabsorbed, only to be secreted once more by the liver. The result is that throughout the digestion and absorption of a meal, the concentration of bile acids in intestinal contents is adequate for their physiological function.

The functions of bile acids in digestion and absorption are the following.

1. Bile acids permit emulsification of fat by reducing the tension of the oil-water interface.
2. Bile acids prevent denaturation of pancreatic lipase as it leaves the surface of emulsified fat droplets.
3. Bile acids, together with 2-monoglycerides, which are the products of fat hydrolysis, form micelles which then dissolve cholesterol, free fatty acids and fat-soluble vitamins.

These functions are discussed in the next section.

During digestion and absorption, bile acids travel witl chyme along the intestine, and they are almost entirely ab sorbed into portal blood. Upon reaching the liver, they ar(again secreted into bile, and consequently they circulate fron liver to intestine to liver and to intestine again. This process i: called the *enterohepatic circulation* of bile acids.

Absorption of bile acids is both active and passive.

Active absorption occurs only in the last part of the ileum and only ionized, conjugated bile acids are actively absorbed The absorption process is so efficient that only a small per centage of conjugated bile acids, less than 5%, reaching the terminal ileum escapes into the colon. In normal man, absorp tion of conjugated bile acids in the ileum accounts for most o the absorptive phase of the enterohepatic circulation.

Passive absorption of bile acids occurs in two ways, botl depending on fat-solubility.

1. Un-ionized conjugated bile acids are more fat-soluble thar are ionized conjugated bile acids. At the pH of intestinal contents, all tauroconjugates are ionized, and they are not passively absorbed. Because glycine is a weaker acid than taurine, a fraction of the glycoconjugates is un-ionized and fat-soluble. A small amount of glycoconjugates is therefore absorbed by passive diffusion through the lipid membrane of the intestinal mucosal cells.

2. Bile acids are attacked by intestinal bacteria. Some bile acids are dehydroxylated; cholic acid loses one hydroxyl group to become deoxycholic acid, and bile acids with two hydroxyl groups lose one to become lithocholic acid. These secondary bile acids, after being absorbed, become part of the bile acid pool. Bacteria also deconjugate acids by hy-drolyzing the peptide bond between the acid and glycine or taurine. Deconjugated bile acids are more fat-soluble than are conjugated bile acids, and therefore some deconju-gated bile acids are passively absorbed. Once they reach the liver, they are reconjugated and secreted in the bile.

As the result of bacterial action, bile acids are deconjugated many times as they circulate. In addition, primary bile acids are converted to secondary bile acids which enter the bile acid pool. One consequence of excessive bacterial action is that the concentration of conjugated bile acids in intestinal contents is reduced, and the concentration of unconjugated bile acids is increased. Since only conjugated bile acids can be actively reabsorbed and since unconjugated bile acids are poorly absorbed only by passive diffusion, a larger fraction of circulating bile acids escapes into the colon. The size of the bile acid pool is reduced, and the fraction of secondary bile acids in the pool is increased.

The rate of enterohepatic circulation is determined by the digestive cycle. During the interdigestive period, most bile acids are sequestered in the gallbladder, and the rate of circulation is low. During digestion of a meal, the gallbladder empties, and bile acids circulate two or more times. In a man eating a low-calorie diet, the pool may circulate 3 to 6 times a day, but on a high-calorie diet, the pool circulates 5 to 14 times a day. If the pool contains a total of 3 gm of bile acids, the man has the use of 9 to 42 gm of bile acids for the purposes of daily digestion.

Although only a small fraction of bile acids fails to be absorbed in the small intestine during each cycle, there are many cycles, and a total of about 500 mg is lost each day. This amount escapes into the colon, where some of it is degraded by bacteria, some absorbed and some excreted in the stool. Bile acids lost are replaced by new bile acids synthesized in the liver, and pool size remains constant. If the terminal ileum is diseased or has been surgically removed, the enterohepatic circulation is broken (Fig 14–4, D). A large fraction of conjugated bile acids is not absorbed but passes into the colon.

The rate at which the liver synthesizes bile acids is governed by the rate at which bile acids return to the liver in portal blood (Fig 14–5). As the rate of return falls, the rate of synthesis rises steeply to a maximum. If about 20% of the bile acids fail to return, synthesis is adequate to keep pool size constant at the normal level. If more than 20% is lost, pool size falls, because the maximum rate of replacement has been reached.

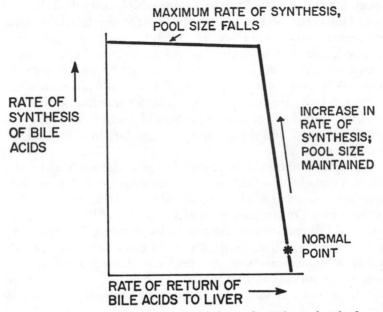

MAXIMUM RATE OF SYNTHESIS,
POOL SIZE FALLS

RATE OF
SYNTHESIS
OF BILE
ACIDS

INCREASE IN
RATE OF
SYNTHESIS;
POOL SIZE
MAINTAINED

NORMAL
POINT

RATE OF RETURN OF
BILE ACIDS TO LIVER

Fig 14–5.—The feedback control of bile acid synthesis by the liver. Under normal circumstances, most of the bile acids return to the liver during each cycle of the enterohepatic circulation. The rate of bile acid synthesis is just sufficient to replace the bile acids lost, and bile acid pool size is maintained. If the rate of return of bile acids to the liver falls, the rate of bile acid synthesis rises sharply. Bile acid pool size remains at the normal level until a little more than 20% of the bile acids is lost during each cycle. Then the maximum rate of synthesis is reached, and with further loss of bile acids the pool size falls.

When the enterohepatic circulation has been broken, there is diurnal variation in the amount of bile acids available. Late at night, when the intestine is empty, the bile acids synthesized by the liver are collected and stored in the gallbladder. As the first meal is eaten, these bile acids are emptied into the duodenum, and they may be adequate for the digestion of the meal. If they are not reabsorbed, however, only the bile acids immediately synthesized are available for the rest of the day. This amount is usually inadequate for normal digestion of fat.

Many compounds precipitate in the gallbladder to form

stones, and although individual stones may be chiefly composed of cholesterol, all stones have several components. Bilirubin is secreted by the liver conjugated with glucuronic acid. If bilirubin is deconjugated in the gallbladder by β-glucuronidase of *E. coli*, its calcium salt precipitates to form pigment stones. Unconjugated bilirubin may also be secreted by the liver and precipitate in the gallbladder.

Cholesterol gallstones form in the gallbladder when minute crystals of cholesterol contained in bile aggregate. Cholesterol is carried in micellar solution in bile secreted by the liver. Since micelles are composed of bile acids, lecithin and cholesterol, the amount of cholesterol that can be carried in micellar solution depends upon the amount of bile acids and lecithin. If the proportion of bile acids and lecithin is high compared with the amount of cholesterol, the cholesterol secreted by the liver is held in micellar solution. When the bile is concentrated in the gallbladder, the proportion of bile acids and lecithin to cholesterol remains unchanged; cholesterol continues to be held in micellar solution. However, if the concentration of bile acids falls below the concentration at which cholesterol can be carried in micelles, bile becomes supersaturated with cholesterol. Microcrystals of cholesterol form in bile, and when bile containing microcrystals is concentrated in the gallbladder, the crystals may coalesce into stones.

In any one person, the rate of secretion of cholesterol into bile is nearly constant over 24 hours. Consequently, the relation between cholesterol in bile and bile acids is determined chiefly by the rate at which bile acids are secreted. During the day when meals are eaten regularly, bile acid secretion is high, and bile is undersaturated with cholesterol. During overnight fasting, bile acid secretion falls, and bile frequently becomes supersaturated with cholesterol. We do not know why supersaturated bile forms gallstones in some persons and not in others.

Persons with cholesterol gallstones have a total bile acid pool smaller than normal persons. Feeding them chenodeoxycholic acid reduces the tendency of the liver to secrete supersaturated bile by expanding the pool and inhibiting cholesterol synthesis in the liver. Unsaturated bile containing a large fraction of chenodeoxycholic acid slowly dissolves cholesterol stones already present in the biliary tract.

About one third of the chenodeoxycholic acid present in the bile acid pool is converted to lithocholic acid each day. Lithocholic acid is sulfated by the liver, and sulfated lithocholic acid is not reabsorbed in the small intestine. Consequently, lithocholic acid is preferentially cleared from the pool, and in normal persons, lithocholic acid is only about 2% of the total bile acid pool. In persons whose bile acid pool has been expanded by exogenous chenodeoxycholic acid, lithocholic acid formation is correspondingly increased, but efficient sulfation of lithocholic acid by the liver prevents its accumulation.

SUGGESTED READINGS

Code, C. F. (ed.): *Handbook of Physiology: The Alimentary Canal* (Washington, D.C.: American Physiological Society, 1968), Chaps. 110–117.

Allan, R. N., Thistle, J. L., and Hofmann, A. F.: Lithocholate metabolism during chenotherapy for gallstone dissolution. 2. Absorption and sulfation, Gut 17:413, 1976.

Cowen, A. E., Korman, M. G., Hofmann, A. F., and Thomas, P. J.: Metabolism of lithocholate in healthy man, Gastroenterology 69: 77, 1975.

Hansen, R. F., and Preis, J. M.: Synthesis and enterohepatic circulation of bile salts, Gastroenterology 73:611, 1977.

Hoffman, N. E., and Hofmann, A. F.: Metabolism of steroid and amino acid moieties of conjugated bile acids in man. V. Equations for the perturbed enterohepatic circulation and their application, Gastroenterology 72:141, 1977.

Northfield, T. C., LaRusso, N. F., Hofmann, A. F., and Thistle, J. L.: Biliary lipid output during three meals and an overnight fast, Gut 16:12, 1975.

Trotman, B. W., Morris, T. A., III, Sanchez, H. M., Soloway, R. D., and Ostrow, J. D.: Pigment versus cholesterol cholelithiasis: Identification and quantification by infrared spectroscopy, Gastroenterology 72:495, 1977.

15. DIGESTION AND ABSORPTION OF FAT

Because fat contains 9 kilocalories per gm, it is a major source of dietary calories. Intake of fat ranges from 12 gm a day (108 kilocalories) in the poorest populations to 150 gm a day (1,350 kilocalories) in the most self-indulgent.

The most important physical property of fat is that fats are poorly soluble in water. For this reason, fat is the major structural component of the body. Fat forms the membranes of cells, and without membranes the body would thaw and resolve itself into a dew.

Insolubility of fat presents a physiological problem: How can fat be transferred from food through the aqueous media of chyme, cell cytoplasm, interstitial fluid, lymph and blood to the organs in which it can be used for energy and structure? To solve this problem, the body carries fat through a complex series of physical and chemical transformations (Fig 15–1). Fat is emulsified and then hydrolyzed to free fatty acids and 2-monoglycerides. These are held in micelles in the lumen until they can be absorbed, and diffusion of micelles carries most of the fat through an unstirred layer to the absorbing surface of the mucosa. Inside the intestinal epithelial cell, most of the free fatty acids and 2-monoglycerides are resynthesized into triglycerides and phospholipids. These are aggregated into small droplets, the chylomicrons, on whose surface a β-lipoprotein is spread, and the chylomicrons are extruded from the cells into the interstitial fluid. They find their way into the lymph and are eventually delivered to the blood. Each step is subject to error, and error at one or another step results in maldigestion or malabsorption of fat.

Triglycerides are the most numerous and important fats in the diet. They are esters of glycerol and three fatty acids. All the fatty acids have an even number of carbon atoms, and most

123

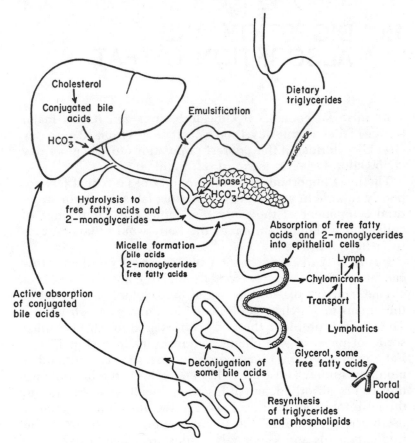

Fig 15–1.—The processes of fat digestion and absorption.

of the fatty acids have long chains, 16 carbon atoms in the case of palmitic acid and 18 in stearic acid. Fatty acids may have one or more double bonds. The fatty acid with 18 carbons and one double bond is oleic acid, and that with two double bonds is linoleic acid.

The longer the chain length of the fatty acids in triglycerides, the higher is the melting point. Triglycerides containing saturated fatty acids have higher melting points than those containing unsaturated acids. The melting point has an important influence upon the digestibility of triglycerides. Triolein is an oil at body temperature, and it is 96% digested and absorbed. On the other hand, tripalmitin is solid at body temperature, and it is only 12% digested and absorbed.

Very few triglycerides containing short-chain fatty acids are eaten in the normal diet. The only common source of short-chain fatty acids is milk and butter, and less than 10% of the fatty acids in butter have a chain length shorter than 10 carbons long. There are almost no medium-length fatty acids 10, 12 or 14 carbons long in the normal diet.

Triglycerides of short-chain and medium-chain fatty acids are more water-soluble than those of long-chain fatty acids, and they can be more completely digested and absorbed when the digestion of triglycerides of long-chain fatty acids is impaired by errors.

A fatty acid esterified at either of the terminal hydroxyl groups of glycerol is said to occupy the 1 or the 1' position. A fatty acid esterified at the middle hydroxyl group is said to occupy the 2 position (Fig 15–2).

Lecithin is a phospholipid; fatty acids are esterified at the 1 and 2 positions of glycerol, and phosphoric acid is esterified at the 1' position (Fig 15–3). An organic base such as choline or ethanolamine is bound in an ester bond to the phosphoric acid. Consequently, the lecithin molecule is polar: the fatty acid end is hydrophobic and the phosphoric acid and base make the other end hydrophilic. There is some lecithin in the normal diet, and there is much lecithin in the bile. Lecithin from both sources is digested and absorbed. The amount of lecithin digested has no influence upon the amount secreted in the bile.

The enzyme phospholipase A_2 is secreted in zymogen form in pancreatic juice, and it is activated in the lumen by trypsin. It catalyzes the hydrolysis of the ester bond of lecithin at the 2 position. The product is lysolecithin: glycerol esterified to a fatty acid at the 1 position and to the phosphoric acid-base compound at the 1' position. The molecule is highly polar, and there is a high concentration of it in duodenal chyme during normal digestion. There is also a phospholipase A_1 in pan-

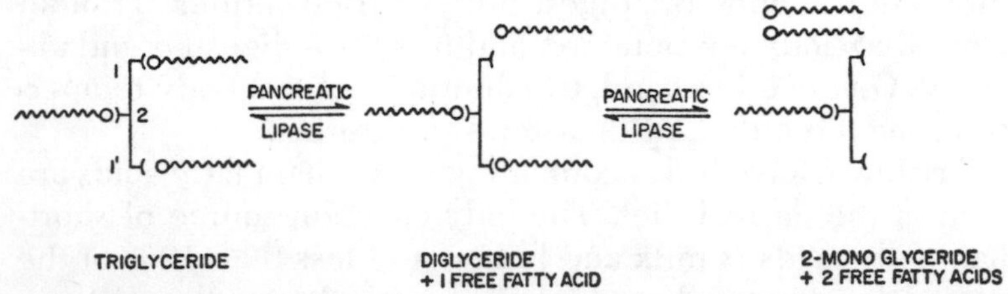

TRIGLYCERIDE   DIGLYCERIDE   2-MONO GLYCERIDE
                          + I FREE FATTY ACID   + 2 FREE FATTY ACIDS

Fig 15–2.—The hydrolysis of a triglyceride catalyzed by pancreatic lipase. The reaction goes completely to the right for the reason that free fatty acids are ionized at the pH of intestinal contents. Pancreatic lipase catalyzes the formation of the ester bond only with un-ionized fatty acids.

creatic juice which catalyzes hydrolysis of the fatty acid ester bond at the 1 position.

The lipase secreted by the gastric mucosa is most active in catalyzing the hydrolysis of triglycerides containing short-chain fatty acids, and it is inactive in an acid medium. It is probably unimportant in adults. Acid in the stomach itself promotes the hydrolysis of some ester bonds, and it tends to break emulsions of fat.

Emulsification of most dietary fat occurs in the duodenum. The process requires a neutral environment and a detergent. The bicarbonate-containing juices of pancreas and liver neu-

Fig 15–3.—The hydrolysis of lecithin of food and bile to lysolecithin and a free fatty acid catalyzed by phospholipase A_2 of pancreatic juice.

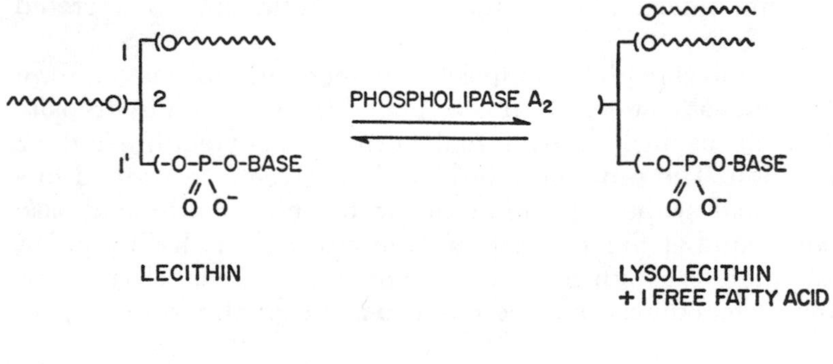

LECITHIN

LYSOLECITHIN
+ I FREE FATTY ACID

tralize acid chyme in the duodenum. Within 20 minutes of the beginning of a meal, the gallbladder begins to contract and empties concentrated bile into the duodenum. Bile acids and lecithin thereupon emulsify fat, forming a stable emulsion of droplets 0.5 to 1 μ in diameter.

Lipase is mixed with duodenal chyme at the same time. Pancreatic lipase has two important properties: (1) it acts as a catalyst only when it is spread on the surface of an emulsified fat droplet with the aid of colipase, and (2) it is almost entirely specific in catalyzing the hydrolysis of the ester bonds at the 1 and 1' positions. Consequently, the products of lipolytic action are free fatty acids and 2-monoglycerides (Fig 15–4). These are the major fat digestion products, which are subsequently absorbed, and during absorption at least 75% of the ester bonds between fatty acids and the hydroxyl groups at the 2 position are preserved intact. These intact ester bonds appear in triglycerides and phospholipids in the chylomicrons of intestinal lymph.

The ester bond in triglycerides is a low-energy bond, and re-esterification is relatively easy. The equilibrium between glycerides and free fatty acids is shifted in the direction of complete hydrolysis by the ionization of free fatty acids at the pH of chyme. Because the free fatty acids are ionized, they cannot reunite with glycerol; only un-ionized fatty acids can esterify.

Another consequence of the fact that the ester bond is a low-energy bond is some hydrolysis at the 2 position. This produces some free glycerol which, being water-soluble, is absorbed by diffusion and passes into the portal blood. Upon reaching the liver, it is metabolized and plays no further part in fat digestion and absorption.

The major products of fat hydrolysis remaining in the chyme are free fatty acids and 2-monoglycerides. These are only slightly soluble in the aqueous medium of chyme. Hydrolysis is so much more rapid than absorption that, unless some additional provision were made, the products of hydrolysis would soon saturate the chyme, and the bulk of the free fatty acids and 2-monoglycerides would separate into an oil or solid

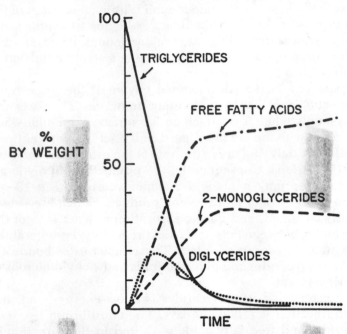

Fig 15–4.—The course of hydrolysis of a triglyceride catalyzed by pancreatic lipase. There is a transient appearance of diglycerides, and the final products of rapid enzymatic hydrolysis are 2-monoglycerides and free fatty acids. Then there is a slow decline in the amount of 2-monoglycerides and a corresponding slow appearance of additional free fatty acids as the ester bond on the 2-position of glycerol is hydrolyzed. At the same time, there is the slow appearance of free glycerol. (After Mattson, F. H.)

phase. Once separated, the free fatty acids and 2-monoglycerides would be essentially unavailable for absorption.

Separation of free fatty acids and 2-monoglycerides into an oil phase is prevented by micelle formation. Micelles are aggregates of fat molecules and bile acids. Fats form the hydrophobic core, and the bile acids, being polar molecules, cover the surface of the micelle, their hydrophobic side facing the core and their hydrophilic side facing the aqueous medium. Each micelle is about 4–6 mμ in diameter. It has a volume a million times smaller than an emulsified fat droplet, and it contains about 20 fat molecules.

The initial constituents of micelles are bile acids and 2-monoglycerides. Two characteristics of bile acids determine whether micelles are formed: their Krafft point and their critical micellar concentration.

The Krafft point is the temperature below which a particular bile acid will not form a micelle. Most bile acids have Krafft points well below body temperature and are therefore capable of forming micelles in the intestinal lumen. The secondary bile acid, lithocholic acid, has a high Krafft point and is incapable of forming micelles at body temperature.

The critical micellar concentration is the minimal concentration of a particular bile acid required for micelle formation (Fig 15–5). When the concentration of the bile acid is at or above its critical micellar concentration, the bile acid and 2-monoglycerides aggregate as micelles, and with increasing bile acid concentration more 2-monoglycerides are carried in micelles. The critical micellar concentration of conjugated bile acids is well below the usual concentration of those bile acids in chyme, and micelles easily form as 2-monoglycerides are produced by the hydrolysis of triglycerides. The critical micellar concentration of unconjugated bile acids is much higher than that of conjugated acids. Consequently, when a considerable fraction of bile acids is deconjugated in the intestinal lumen by bacterial action, micelle formation is reduced. When deconjugated bile acids predominate, fewer conjugated bile acids are present to form micelles, and the deconjugated bile acids cannot form micelles on account of their high critical micellar concentration.

Once micelles are formed by bile acids and 2-monoglycerides, they dissolve other fat-soluble compounds. Quantitatively, the most important of these are the free fatty acids. Fat-soluble vitamins are carried in micelles, and this is the reason bile acids are essential for absorption of vitamin K. Cholesterol and esters of cholesterol, compounds which are almost totally insoluble in water, are dissolved in micelles.

The constituent molecules of micelles move back and forth

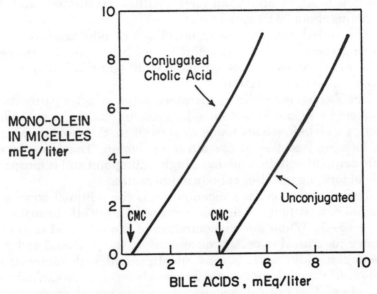

Fig 15–5.—Relationship between bile acid concentration and the quantity of mono-olein brought into micellar solution at pH 6.3, 37°C and [Na+] of 150 mN. The critical micellar concentration (CMC) of the conjugated cholic acid is 0.25 mM, a concentration well below that usually occurring in intestinal contents. The CMC for the unconjugated bile acid is 4 mM. (Adapted from Hofmann, A. F., and Borgstrom, B.: Fed. Proc. 21:43, 1962.)

between micelles and solution with great rapidity. The mean residence time of a particular molecule in a micelle is only 10 milliseconds. This means that as free fatty acids and 2-monoglycerides are absorbed from solution into the epithelial cells, the aqueous phase of chyme is kept saturated by movement of free fatty acids and 2-monoglycerides from micelles to solution. Micelle formation prevents the separation of the products of lipolysis into an oil phase, and micelles serve as a reservoir of lipolytic products so that the aqueous phase is always saturated.

Fat is digested in the bulk phase of the intestinal lumen where the contents are well stirred. Fat digestion products are absorbed through the apical membrane of mucosal cells, and that membrane is covered by a thin unstirred layer through which the fat digestion products must pass to be absorbed.

The solution is kept saturated by exchange with micelles, but even at saturation, the concentration of fat digestion products in solution is very low. Consequently, only a small quantity of free fatty acids and 2-monoglycerides in solution can diffuse through the unstirred layer. This quantity is about 1% of the total absorbed. Micelles, being larger, diffuse more slowly, but because their concentration is high, they carry the remaining 99% of free fatty acids and 2-monoglycerides through the unstirred layer to the cell membrane (Fig 15-6).

Most free fatty acids and 2-monoglycerides are absorbed in the duodenum and upper jejunum. Most bile acids are absorbed in the terminal ileum. A little cholesterol is absorbed throughout the length of the intestine, but most escapes into the colon. Therefore, the constituents of micelles are separated as the chyme passes down the intestine; their content of free fatty acids and 2-monoglycerides diminishes and their concentration of cholesterol rises.

Once inside the intestinal epithelial cells, the 2-monoglycerides undergo little or no further hydrolysis (Fig 15-7). They are resynthesized into triglycerides. Some of the free fatty acids are combined with glycerophosphate derived from glucose to form phospholipids, and some of the free fatty acids escape into the portal blood to reach the liver. The fraction of free fatty acids going directly to the liver depends upon their chain length. No more than 15% of the long-chain fatty acids goes to the liver in portal blood, but almost all the short-chain acids do.

The intestinal epithelial cells collect the newly synthesized triglycerides and phospholipids, together with some cholesterol, into droplets called chylomicrons. About 10% of the surface of the chylomicrons is covered with a β-lipoprotein synthesized by the epithelial cells. If the cells are incapable of synthesizing this protein, only large chylomicrons, or none at all, are formed. The cells extrude the chylomicrons from their lateral borders into the interstitial fluid.

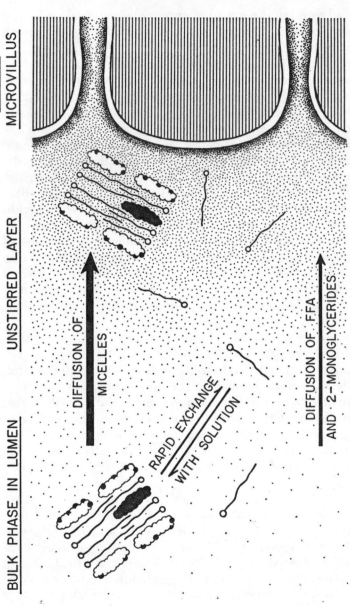

Fig 15–6.—The role of micelles in fat absorption: keeping the solution saturated with free fatty acids (*FFA*) and 2-monoglycerides and carrying fat digestion products through the unstirred layer.

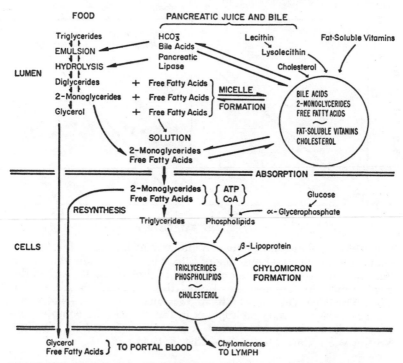

Fig 15-7.—The process of fat digestion and absorption. It is possible that for some substances such as cholesterol to be absorbed the micelles must actually touch the membrane of the microvilli of the intestinal epithelial cells.

Chylomicrons are absorbed into the intestinal lymphatic vessels and transported through the thoracic duct into the blood. Chylomicrons enter the lymphatic vessels rather than the intestinal capillaries, because the fenestrations of the lymphatic vessels are open whereas those of the capillaries are closed by a basement membrane.

Under normal circumstances, there is less than 6 gm of fat in the stool each day, and the amount of fat in the stool is independent of the diet. Most normal fecal fat is contained in bacterial cells. If there are errors in fat digestion and absorption,

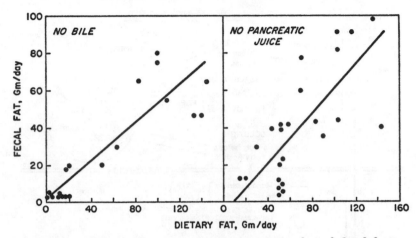

Fig 15–8.—Relationship between dietary fat and total fecal fat in man. *Left*, in complete absence of bile. The slope of the regression line indicates that on the average 46% of dietary fat appears in the stool in the complete absence of bile. *Right*, in complete absence of pancreatic juice. The slope of the regression line indicates that on the average 68% of dietary fat appears in the stool in the complete absence of pancreatic juice. (Adapted from Annergers, J. H.: Q. Bull. Northwestern Univ. Med. School 23:198, 1949.)

fat escapes into the stool. Excess fat in the stool, or steatorrhea, is usually defined as more than 6 gm a day.

In steatorrhea, the amount of fecal fat is roughly proportional to the amount of dietary fat (Fig 15–8), but the fat in the stool is different in some respects from the fat in the diet. Most fecal fat is in the form of free fatty acids; triglycerides of the diet are hydrolyzed and the glycerol absorbed. All the short-chain and most of the medium-chain fatty acids of the diet are absent, for they too are absorbed. Abnormal fats are present. Intestinal bacteria add hydroxyl groups to the double bonds of unsaturated fatty acids. One of the products is ricinoleic acid, the active ingredient of castor oil, and the presence of this hydroxylated fatty acid accounts, in part, for the diarrhea which often accompanies steatorrhea.

Errors in fat digestion and absorption leading to steatorrhea are summarized in Table 15–1.

TABLE 15-1.—ERRORS IN FAT DIGESTION AND ABSORPTION LEADING TO STEATORRHEA

STEP	PHYSIOLOGICAL DISTURBANCE	DISEASE STATE
Emulsification of triglycerides	Impaired emulsification	Deficiency of conjugated bile acids; excessive acidity of intestinal chyme
Hydrolysis	Pancreatic lipase deficiency	Pancreatic disease
	Absolute or relative bicarbonate deficiency	Pancreatic disease or gastric hypersecretion
Formation of micelles	Conjugated bile acid deficiency; absolute or relative bicarbonate deficiency	Biliary fistula or obstruction; ileal disease or resection; bacterial deconjugation; pancreatic disease or gastric hypersecretion
Absorption of 2-monoglycerides	Decreased cell uptake; reduction in cell number, activity or surface area	Intestinal resection or bypass; tropical sprue or gluten enteropathy
	Cells saturated with fatty acids and monoglycerides	Failure of triglyceride synthesis, chylomicron formation or transport
	Decreased contact time	Increased transit time
Chylomicron formation	Deficiency in chylomicron formation	A-β-lipoproteinemia
Transport of chylomicrons from cells via lymph to blood	Lymphatic obstruction or lymphangiectasia	Lymphosarcoma; intestinal lipodystrophy; protein-losing enteropathy

The remarkable fact is that if 40–60% of dietary fat fails to be absorbed, then 60–40% of dietary fat actually is absorbed in diseased states. Processes responsible are poorly understood.

Medium-chain triglycerides are not abundant in nature, but medium-chain triglycerides commercially synthesized are often used to supply calories to persons with errors in fat digestion and absorption. These triglycerides, containing fatty acids 6 to 12 carbons long, are rapidly and completely hydrolyzed in the intestinal lumen, and they can also be hydrolyzed within the epithelial cells. Hydrolysis and micelle formation, however, are not necessary for their absorption. Medium-chain triglycerides which are hydrolyzed are not reconstituted in the epithelial cells; their component fatty acids and glycerol are carried in the portal blood to the liver.

Cholesterol and cholesterol esters of the diet are only a small fraction of the cholesterol handled by the intestine. A large amount of cholesterol is contained in the bile. Intestinal epithelial cells synthesize cholesterol, and the cholesterol they contain also enters the lumen when the cells are desquamated.

Only cholesterol contained in micelles is capable of being absorbed; the rest is too insoluble. It is possible that absorption of cholesterol occurs only when cholesterol-containing micelles touch the surface of the microvilli of intestinal epithelial cells. Then the cholesterol is transferred to the lipid layer encasing the microvilli.

SUGGESTED READINGS

Code, C. F. (ed.): *Handbook of Physiology: The Alimentary Canal* (Washington, D. C.: American Physiological Society, 1968), Chaps. 70, 71, 72.

Poley, J. R., and Hofmann, A. F.: Role of fat maldigestion in pathogenesis of steatorrhea in ileal resection. Fat digestion after two sequential test meals with and without cholestyramine, Gastroenterology 71:38, 1976.

Romell, K., and Goebell, H. (eds.): *Lipid Absorption: Biochemical and Clinical Aspects* (Baltimore: University Park Press, 1976).

Schmitt, M. G., Soergel, K. H., and Wood, C. M.: Absorption of short chain fatty acids from the human jejunum, Gastroenterology 70: 211, 1976.

Wilson, F. A., and Dietschy, J. M.: Characterization of bile acid absorption across the unstirred layer and brush border of the rat jejunum, J. Clin. Invest. 51:3015, 1972.

Wiseman, G.: *Absorption from the Intestine* (New York: Academic Press, 1964).

16. ABSORPTION OF VITAMINS

Vitamins are a heterogeneous class of compounds united only by the body's absolute need for them (Table 16–1).

Fat-soluble vitamins are absorbed like other fats. In the lumen of the intestine, they dissolve in micelles composed of 2-monoglycerides, free fatty acids and bile salts, and micelles carry the vitamins through the unstirred layer to the mucosal surface. Then the vitamins are absorbed by passive diffusion through the lipid membrane of the epithelial cells. Active transport for fat-soluble vitamins has not been described.

When bile salts are absent, micelle formation is defective, and fat-soluble vitamins are poorly absorbed. This is particularly serious in the case of vitamin K, which is required for prothrombin synthesis in the liver. A patient in whom bile salt delivery to the intestine has been interrupted is in danger of serious hemorrhage.

The molecular weight of water-soluble vitamins ranges from 122 (nicotinamide) to 1,355 (cyanocobalamin). The smallest of them can be absorbed by passive diffusion through water-filled pores in the mucosal membrane. Ascorbic acid, being closely related to hexoses, is absorbed by a Na^+-dependent process. Vitamins of intermediate molecular weight are absorbed by specific active transport systems. The one responsible for absorption of thiamine and thiamine pyrophosphate is saturated when the concentration of the vitamin is 1.5 μM; at higher concentration additional vitamin is absorbed by diffusion. Although both thiamine and thiamine pyrophosphate are absorbed into the epithelial cells, only free thiamine is liberated into the blood.

Folic acid occurs naturally conjugated with several glutamic acid residues, but it is hydrolyzed to monoglutamyl folate before it is actively absorbed.

TABLE 16-1.—ABSORPTION OF VITAMINS

FAT-SOLUBLE VITAMINS, solution in micelles required for absorption:		
A	Retinol (MW 286), carrotine (MW 537)	Saturable, carrier-mediated diffusion
D	Calciferol (MW 397) and congeners	Passive diffusion
E	α-Tocopherol (MW 431)	Passive diffusion
K	Vitamin K (MW 451), menadione (MW 172)	Passive diffusion
WATER-SOLUBLE VITAMINS:		
B_1	Thiamine (MW 301), thiamine pyrophosphate (MW 479)	Active transport, plus diffusion
B_2	Riboflavin (MW 376)	Diffusion only (?)
B_3	Nicotinic acid (MW 123), nicotinamide (MW 122)	Diffusion only
B_5	Pantothenic acid (MW 219)	Diffusion only
B_6	Pyridoxal (MW 167) and congeners	Diffusion only
B_{12}	Cyanocobalamin (MW 1,355) and conjugates	Combines with intrinsic factor from stomach, absorbed by pinocytosis in terminal ileum. In absence of intrinsic factor, absorbed in microgram amounts when fed in milligrams.
B_c	Folic acid (MW 441) and conjugates	Conjugates hydrolyzed to mono-glutamyl-folate before active absorption
C	Ascorbic acid (MW 176)	Absorbed by Na^+-dependent active transport, plus diffusion
H	Biotin (MW 244)	Active transport

Cyanocobalamin (vitamin B_{12}) has a molecular weight of 1,355, and its conjugates are even larger. They are too big to be absorbed by diffusion. The gastric mucosa secretes a muco-protein, intrinsic factor, which combines with cyanocobala-min in the upper small intestine. In its journey to the terminal ileum, the complex of cyanocobalamin and intrinsic factor may be attacked by bacteria or by intestinal parasites which need the vitamin for themselves. The complex which reaches the terminal ileum attaches to specific receptor sites on the epithelial cells, and the complex is absorbed by pinocytosis.

The vitamin is then slowly passed on to the blood by a succession of carriers.

Vitamin B_{12} is required in microgram amounts for maturation of erythrocytes. In a person whose atrophic gastric mucosa does not secrete intrinsic factor, absorption by the process described cannot occur, and unless he is treated, such a person dies of pernicious anemia. Before the availability of pure cyanocobalamin for parenteral administration, a patient with pernicious anemia was fed a half pound or more of raw liver a day. That amount of liver contains milligrams of the vitamin, and the few micrograms required could be absorbed without intrinsic factor.

Absorption of intrinsic factor combined with cyanocobalamin occurs only in the terminal ileum. A person whose terminal ileum has been removed develops pernicious anemia about 4 years after the operation when the supply of vitamin stored in his liver is exhausted.

Some vitamins, thiamine and vitamin K for example, are synthesized by microflora inhabiting the gut. In some animals such synthesis may be the only source of one or another vitamin, but in man this is probably not important.

SUGGESTED READINGS

Code, C. F. (ed.): *Handbook of Physiology: The Alimentary Canal* (Washington, D. C.: American Physiological Society, 1968), Chaps. 51, 77.

Glass, G. B. J.: Gastric intrinsic factor and its function in the metabolism of vitamin B_{12}, Physiol. Rev. 43:529, 1963.

Rindi, G., and Ventura, U.: Thiamine intestinal transport, Physiol. Rev. 52:821, 1972.

Rosenberg, I. H., and Godwin, H. A.: The digestion and absorption of dietary folate, Gastroenterology 60:445, 1971.

Toskes, P. P., and Deren, J. J.: Vitamin B_{12} absorption and malabsorption, Gastroenterology 65:662, 1973.

Wiseman, G.: *Absorption from the Intestine* (New York: Academic Press, 1964).

17. GAS IN THE GUT

The four sources of intestinal gas are swallowing, fermentation, neutralization and diffusion.

1. *Swallowing.* Air is swallowed with food and drink in very variable amounts. Some persons swallow so much while eating that they become uncomfortably bloated, whereas the air swallowed by others accounts for less than a third of the gas in the gut. Air is also swallowed in frothy saliva, and much may be swallowed during copious salivation accompanying nausea. Oxygen is removed from swallowed air by the flora of the gut, and the residual nitrogen is excreted as flatus.

2. *Fermentation.* Under normal circumstances, there are few bacteria in the small intestine. Gastric contents are partially sterilized by acid secreted by the stomach, and chyme moves rapidly from duodenum to colon. When chyme moves with abnormal slowness or when food residues are sequestered in blind loops or diverticuli, bacteria multiply. Then the small intestine becomes bloated with gas that is produced by fermentation. Fermentation and gas production normally occur in the slowly moving contents of the colon. Colonic flora of all persons produce variable amounts of hydrogen gas. About 14% is absorbed and excreted through the lungs; the rest escapes in flatus. About a third of the population produces a large amount of methane in the colon, but the other two-thirds does not. Some methane is also absorbed into the blood, and its mass-spectrometric measurement in breath is a means by which its production in the gut can be followed. Methane, together with hydrogen, makes the flatus combustible. Hydrogen sulfide is only a small fraction of a percent of the gas passed as flatus.

143

The amount of hydrogen and methane produced depends upon the nature of the unabsorbed residue. The hulls of beans contain oligosaccharides which cannot be hydrolyzed by intestinal oligosaccharidases, and when beans are eaten to the extent of 25% of the caloric intake, the rate of passage of flatus may be as high as 200 cc per hour. Deficiency of lactase leaves lactose to be fermented by intestinal flora, and when milk is drunk by a person with lactase deficiency, he makes as much as 4 cc of gas a minute. All varieties of fermentation produce organic acids that are partially neutralized in the small intestine or colon.

3. *Neutralization.* When 1 mEq of acid reacts with 1 mEq of bicarbonate at body temperature, 25 cc of carbon dioxide is liberated. During emptying of a meal from the stomach, 50 mEq of acid from the stomach may react with an equivalent amount of bicarbonate from the pancreas, and 1,250 cc of carbon dioxide bubbles out of duodenal contents. The partial pressure of carbon dioxide rises as high as 700 mm Hg, and most of the carbon dioxide resulting from neutralization in the small intestine diffuses back into the blood. Carbon dioxide resulting from reaction of organic acids formed in the colon with bicarbonate secreted by the colonic mucosa is more slowly absorbed, and after beans are eaten the mole fraction of carbon dioxide in flatus is as high as 0.62.

4. *Diffusion.* Addition of other gases to nitrogen in the gut reduces the partial pressure of nitrogen, and then nitrogen diffuses from blood into the gas phase in the gut at a rate of 1–2 cc a minute. Any oxygen diffusing from blood is used by colonic flora.

In man in the upright position, a gas bubble is trapped in the fundus of the stomach above the entrance of the esophagus. A characteristic volume for this bubble is 50 cc. There is another 100 cc of gas in transit through the small intestine and colon, making a total of 150 cc.

The stomach can be filled with a large volume of gas before belching, or eructation, occurs. When the gas bubble in the stomach is large enough to extend below the entrance of the esophagus, gas may escape into the esophagus when the lower esophageal sphincter relaxes, but it does not pass the hypopharyngeal sphincter. Distention of the esophagus by gas stimulates secondary peristalsis, and the gas is swept back into the stomach. This sequence of filling and emptying of the esophagus may repeat many times before eructation occurs. Then, when the esophagus is filled with gas, the jaw is thrust forward, the abdominal muscles contract, expiration is attempted against a closed glottis and gas is expelled through the partially opened hypopharyngeal sphincter.

Gas moves rapidly through the intestinal tract, because it has low viscosity. A mixture of gas and liquid squirting through narrow parts of the gut produces sounds with a wide range of pitch; a low rumble is call a *borborygmus*. A single segmental contraction in the small intestine or colon pushes a bubble of gas forward through several segments, whereas it pushes much more viscous chyme only a short distance. This difference in viscosity accounts for the fact that gas can be expelled through the anus without accompanying feces, for by the time the extremely viscous feces have moved, wind has been broken and the external anal sphincter has tightly contracted again. The external anal sphincter relaxes reflexly during micturition, and this accounts for the frequent passing of flatus while urinating.

It is not the fat content but the entrained gas bubbles that cause feces to float, and feces may be made to sink or rise like a Cartesian diver by increasing or decreasing the ambient pressure.

SUGGESTED READINGS

Code, C. F. (ed.): *Handbook of Physiology: The Alimentary Canal* (Washington, D.C.: American Physiological Society, 1968), Chap. 137.

Berk, J. E. (ed.): Gastrointestinal gas, New York Acad. Sci. 150:1, 1968.

Levitt, M. D., and Bond, J. H., Jr.: Volume, composition, and source of intestinal gas, Gastroenterology 59:921, 1970.

Politzer, J-P., Devroede, C., Vasseur, C., Gerard, J., and Thibault, R.: The genesis of bowel sounds: Influence of viscus and gastrointestinal content, Gastroenterology 71:282, 1976.

18. THE COLON

The parts of the colon and their names are shown in Fig 18-1.

The intrinsic plexuses of the colon are similar to the plexuses of the small intestine, and they have a similar important role in regulating movements of the colon. The vagus nerve provides parasympathetic innervation to the proximal two-thirds of the colon. Parasympathetic innervation to the distal colon, the rectum and the internal anal sphincter comes from sacral segments of the spinal cord, and it is particularly important in governing defecation. Muscle at the anus is the external anal sphincter, which is composed of striated muscle, and therefore the behavior of the external anal sphincter is entirely controlled by motor neurons, the cell bodies of which are in the sacral segments of the cord. Sympathetic innervation of the colon acts to constrict blood vessels, and it is partly excitatory, partly inhibitory for colonic smooth muscle.

In general, movements of the colon are very slow, and the first radiologist to observe the colon said that it presented a picture of still life. Slow movements of its contents are appropriate for the colon's function of absorbing water and electrolytes. The colon receives a total of about 1,500 ml a day, but this is delivered at a very irregular rate. During digestion of a meal, fluid may pass through the ileocecal valve at the rate of 15 ml a minute, but none may enter the colon for a long time during the interdigestive state. The colonic mucosa actively reabsorbs Na^+ and Cl^-, and water is absorbed along with electrolytes. The colonic mucosa also secretes K^+ and HCO_3^- into its contents. Absorption of Na^+ and secretion of K^+ is partially controlled by aldosterone, and when aldosterone concentration in the plasma is elevated, absorption of Na^+ and secretion of K^+ are enhanced. Acids produced by fermentation in the colon are partially neutralized by secreted HCO_3^-. The ab-

147

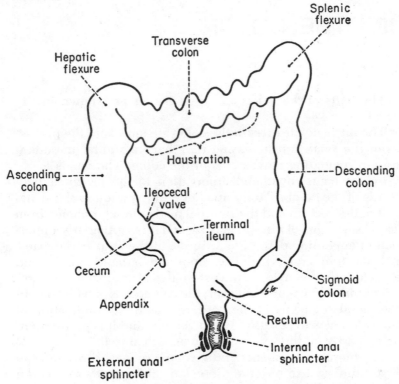

Fig 18–1.—The parts of the colon.

sorptive capacity of the colon is at least 2,000 ml a day when fluid is delivered to it in a steady stream. When the colon's capacity to absorb is exceeded, either by a sudden gush of fluid or by a large amount entering slowly, diarrhea results. Absorption of Na^+, and consequently of water, in the colon is inhibited by some bile acids, and therefore failure of absorption of bile acids in the terminal ileum can also be a cause of diarrhea.

Regularly spaced, ringlike contractions of the circular muscle of the colon divide it into *haustra*. Slow contraction of one ring of circular muscle is replaced by slow relaxation, and the neighboring, hitherto relaxed ring of circular muscle slowly contracts. This behavior of the circular muscle of the colon is similar to that causing segmentation in the small intestine, with the difference that colonic contractions and relaxations occur over many minutes, whereas those of the small intestine occur in seconds.

Much of the time the contents of the haustra are shuttled back and forth with no progressive movement. About one third of the time during the interdigestive period, haustral contents are moved in both directions, orthograde and retrograde with orthograde movement being slightly more frequent. Thus the contents are moved very slowly as absorption of water and electrolytes takes place.

Occasionally multihaustral segmental propulsion such as that shown in Figure 18–2 occurs. This type of movement displaces a substantial bulk of colonic contents in the orthograde direction.

Less frequently peristalsis consisting of a progressive wave of contraction preceded by a progressive wave of relaxation pushes colonic contents forward at a rate of 1–2 cm a minute. Reverse peristalsis may occur, but it is extremely rare in man.

In normal persons, between meals these several kinds of movements of the colon push the contents forward at the rate of about 8 cm an hour and backward at the rate of 3 cm an hour. Net forward movement is 5 cm an hour. In constipated persons, net forward movement is only 1 cm an hour. Slow and reluctant is the long descent with many a lingering farewell look behind.

After a meal, the colon is aroused from its torpor. Haustral shuttling to no effect decreases, and orthograde propulsion by segmentation increases. The frequency of multihaustral segmental propulsion doubles, and peristalsis is slightly more frequent. The result is that after eating, forward movement averages 14 cm an hour, backward movement 3 cm an hour and net forward movement 11 cm an hour. Administration of a long-lived cholinergic drug increases net forward movement to 20 cm an hour.

Increased motility of the colon after meals frequently pushes feces into the rectum and arouses the urge to defecate. Dis-

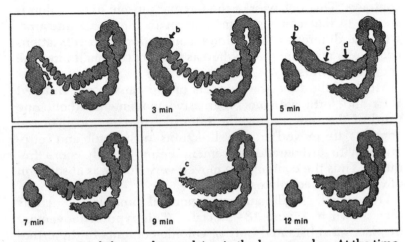

Fig 18–2. — Multihaustral propulsion in the human colon. At the time of the first observation *(top left)*, barium-impregnated ileal contents *(arrow a)* were entering the cecum. Contraction of the cecum, 3 minutes later, propelled a large part of its contents upward to distend the region of the hepatic flexure *(arrow b)*. After another 2 minutes, this section had also contracted, and its contents were distributed over the proximal half of the transverse colon. Haustral markings disappeared over most of this length, and there was some narrowing of the mass in 2 to 3 inches of the middle section *(arrow c)*. Material forced out of the narrow section was accommodated by distention of the next four haustra *(arrow d)*. As haustration was returning, 2 minutes later, most of the proximal half of the transverse colon between *b* and *c* also contracted. All the contents expelled from this section lodged in the distal half of the transverse colon. The conical outline at *c* in the fifth picture is typical of multihaustral contraction. (Drawings made from cineradiographic studies; adapted from Ritchie, H. A.: Gut 9:442, 1968.)

tention of the rectum is the primary stimulus for defecation. The basic mechanism of the defecation reflex resides in the intramural plexuses of the colon and the muscles of the rectum and internal anal sphincter. When the rectum is distended, its walls are stretched, and pressure in the rectum rises (Fig 18–3). If the stretch is small or brief, there may be no reflex contraction of the rectum, and tension in the wall and pressure in the lumen may return to baseline levels. If stretch is greater, receptors in the wall of the rectum are stimulated, and, through a reflex are confined to the intramural plexuses, the rectum is stimulated to contract. Pressure in the rectum then rises further. At the same time the internal anal sphincter, which is a thickening of the smooth muscle at the end of the rectum, relaxes. Contraction of the rectum and relaxation of the sphincter tend to expel rectal contents. This behavior takes place when there is no extrinsic innervation of the rec-

Fig 18–3.—Response of the rectum, internal anal sphincter and external anal sphincter to distention of the rectum. Some distention of the rectum stretches its wall and causes a passive increase in pressure. More distention causes more passive increase in pressure and is followed by further increase in pressure resulting from active contraction. Still more distention is followed by greater active contraction, and subsequent contractions may occur rhythmically at about 20-second intervals. Each increase in pressure in the rectum is accompanied by a decrease in pressure in the internal anal sphincter and an increase in pressure in the external anal sphincter. (Rectal and internal anal sphincter pressures adapted from Denny-Brown, D., and Robertson, E. G.: Brain 58:256, 1935; external anal sphincter pressures adapted from Schuster, M. M., et al.: Bull. Johns Hopkins Hosp. 116:79, 1965.)

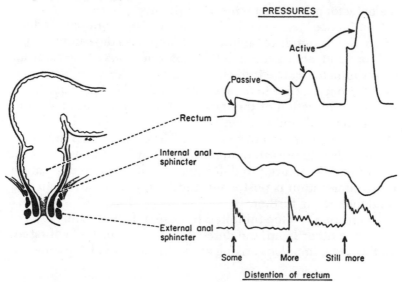

tum and internal anal sphincter, and there is, therefore, no possiblity that the defecation reflex can be modified through extrinsic reflexes.

Under normal circumstances, however, this basic reflex is modified by reflex activity in the neuraxis. The reflex can be enhanced so that defecation occurs, or it can be inhibited.

The external anal sphincter is a ring of striated muscle partly surrounding the internal anal sphincter and closing the anal canal. This sphincter is kept in a state of tonic contraction by reflexes beginning in its own muscle spindles. The spindles send impulses to the sacral part of the spinal cord, and corresponding impulses return to the external anal sphincter through motor neurons controlled, in part, by afferent impulses from the spindles. The strength of contraction of the external anal sphincter can be increased or decreased by other reflex influences converging on the motor neurons.

When the rectum is distended, the internal anal sphincter relaxes, but the external anal sphincter contracts (Fig 18–3). The stimulus for contraction is activation of stretch receptors in the wall of the rectum, the afferent fibers of which go to the spinal cord. This contraction of the external anal sphincter is a major factor providing for fecal continence.

Continence can usually be ensured by voluntary effort. When defecation is to be inhibited, impulses descend the spinal cord and, acting through the motor neurons to the external anal sphincter, cause it to close tightly. At the same time, impulses from the cord to the rectum and distal colon cause them to relax. Relaxation of the muscle of the rectum reduces tension in its wall, and although the rectum may be full of feces, stretch receptors are not activated. They no longer send afferent impulses to the intramural plexuses and to the spinal cord. There is now no urge to defecate, and there is no defecation reflex; defecation is postponed until more feces arrive to distend the rectum further.

Defecation is often inhibited by pain or fear of pain.

On the other hand, the defecation reflex can be facilitated. In this case, impulses arising at the highest level converge on a "defecation center" in the medulla, and from there facilitating impulses pass to the sacral section of the cord. The external anal sphincter and other perineal muscles relax, and the anal canal is partly extruded. Contraction of the longitudinal muscles of the rectum shortens the rectum and obliterates the angle the distal colon makes with the rectum. Peristaltic waves sweep from the sigmoid colon to the rectum, pushing feces through the relaxed internal and external anal sphincters at a rate ranging from stately slowness to explosive violence. At the end of defecation, there is rebound contraction of the external anal sphincter.

In a normal person, evacuation is assisted by a large increase in intra-abdominal pressure brought about by contractions of the chest muscles on a closed glottis and simultaneous contraction of abdominal muscles. The hemodynamic consequences of this maneuver—the Valsalva maneuver—are an abrupt rise in arterial pressure as the increase in intrathoracic pressure is transmitted across the wall of the heart, stoppage of venous return with subsequent fall in cardiac output and then a fall in arterial pressure. Death while straining at stool results from cerebrovascular accidents, ventricular fibrillation or coronary occlusion. Consequently, the appropriate place to look for an elderly person who does not answer the telephone is on the floor of the bathroom.

In some persons, particularly children, the ganglia of the intrinsic plexuses of a part of the distal colon or rectum degenerate for unknown reasons. Then the defecation reflex fails, and the colon above the affected part becomes enormously distended. This condition is called *megacolon*. If the aganglionic segment can be identified and removed, with anastomosis of the normal part of the colon above to the normal part below, the colon can empty regularly, and the distention disappears.

Psychosomatic factors, from the most deeply buried to the most overt, influence functions of the colon. The reaction of extreme fright and the use of the colon to express hostility or disdain are familiar to all. The irritable or spastic colon is characterized by abnormal motor function, abdominal pain and flatulence. Constipation alternates with diarrhea, and the condition waxes and wanes in step with changes in the life situation. Medical students need not be reminded of cycles of diarrhea in phase with examinations.

The normal natural frequency of defecation ranges from after every meal to once every 3 days or so. Stool weight is strongly influenced by the amount of undigestible fiber in the diet. The weight of individual stools of a person on the average American diet is 100–125 gm, but for Ugandans who eat a large amount of cereal bran, the weight may be as great as 980 gm.

It is difficult to specify what infrequency of defecation constitutes constipation, but when defecation is unduly prolonged, the victim experiences mental depression, restlessness, dull headache, loss of appetite sometimes accompanied by nausea, a foul breath and coated tongue and abdominal distention. This miserable state is exacerbated by fear of its consequences, and many persons are convinced that unless the colon is cleaned out regularly, toxins absorbed from the colon will poison them. They resort to colonic irrigation and heroic doses of laxatives.

Some toxic compounds are produced by bacterial decarboxylation of amino acids, but if these are absorbed, they are detoxified by the liver. Urea is hydrolyzed to ammonia by urease in the colon. Ammonia (NH_3) is absorbed by passive diffusion, but ammonium (NH_4^+) is not. In a normal person, any ammonia absorbed into portal blood is at once converted to urea in the liver, but in a patient with portocaval shunt, ammonia escapes into the systemic circulation and may cause fatal encephalopathy. In such a patient, ammonia production may be reduced by treatment with antibiotics. Ammonia absorption may be decreased by feeding lactulose (1,4-galactosidofructose) for which there is no naturally occurring hydrolytic enzyme in the intestinal mucosa. The unabsorbed lactulose is fermented by bacteria in the colon, and the acid produced converts ammonia to unabsorbable ammonium. Lactulose also reduces ammonia absorption by stimulating the incorporation of ammonia into bacterial proteins.

SUGGESTED READINGS

Code, C. F. (ed.): *Handbook of Physiology: The Alimentary Canal* (Washington, D. C.: American Physiological Society, 1968), Chaps. 101, 102, 103, 134, 136.

Burkitt, D. P., Walker, A. R. P., and Painter, N. S.: Effect of dietary fibre on stools and transit time, and its role in the causation of disease, Lancet 2:1408, 1972.

Cummings, J. H.: Laxative abuse, Gut 15:758, 1974.

Daniel, E. E., Bennett, A., Misiewicz, J. J., Edmonds, C. J., Hill, M. J., and Cummings, J. H.: Symposium on colon function, Gut 16: 298, 1975.

Phillips, S. F., and Geller, J.: The contribution of the colon to electrolyte and water conservation in man, J. Lab. Clin. Med. 81:733, 1973.

Schuster, M. M.: The riddle of the sphincters, Gastroenterology 69: 249, 1975.

Vaughan, A. M., and Martin, J. A.: Foreign body (case knife) in the sigmoid, J.A.M.A. 130:29, 1940.

Wiggins, H. S., and Cummings, J. H.: Evidence for the mixing of residue in the human gut, Gut 17:1007, 1976.

Young, S. J., Alpers, D. H., Norland, C. C., and Woodruff, R. A., Jr.: Psychiatric illness and the irritable bowel syndrome, Gastroenterology 70:162, 1976.

INDEX

157